The Anti-Nuclear Power Movement and Discourses of Energy Justice

The Anti-Nuclear Power Movement and Discourses of Energy Justice

Jesse P. Van Gerven

LEXINGTON BOOKS
Lanham • Boulder • New York • London

Published by Lexington Books
An imprint of The Rowman & Littlefield Publishing Group, Inc.
4501 Forbes Boulevard, Suite 200, Lanham, Maryland 20706
www.rowman.com

86-90 Paul Street, London EC2A 4NE, United Kingdom

British Library Cataloguing in Publication Information Available

Library of Congress Cataloging-in-Publication Data

Names: Van Gerven, Jesse P., 1980- author.
Title: The anti-nuclear power movement and discourses of energy justice / Jesse P. Van Gerven.
Description: Lanham : Lexington Books, [2022] | Includes bibliographical references. | Summary: "This study analyzes anti-nuclear power organizations' claims regarding public financing for new nuclear construction, issues associated with the management of high-level radioactive waste, and other campaigns to increase the safety of nuclear facilities. This leads the author to the identification of general principals of energy justice"—Provided by publisher.
Identifiers: LCCN 2021053800 (print) | LCCN 2021053801 (ebook) | ISBN 9781793620453 (cloth ; alk. paper) | ISBN 9781793620460 (epub)
Subjects: LCSH: Antinuclear movement—United States. | Nuclear industry—Environmental aspects—United States. | Nuclear power plants—Environmental aspects—United States.
Classification: LCC HD9698.U52 V36 2022 (print) | LCC HD9698.U52 (ebook) | DDC 333.792/40973—dc23/eng/20211104
LC record available at https://lccn.loc.gov/2021053800
LC ebook record available at https://lccn.loc.gov/2021053801

Dedication

To Eldon and Sterling Van Gerven. May your lives be filled with love and laughs.

Contents

List of Figures and Tables ix

Acknowledgments xi

Introduction 1

1 Federal Financing of New Nuclear Reactor Construction 35

2 High-Level Radioactive Waste Management 57

3 Local Financing of New Nuclear Reactor Construction and Increased Nuclear Safety 93

Conclusion 117

References 143

Index 151

About the Author 157

List of Figures and Tables

FIGURES

Figure 0.1	Diagram of a Uranium Nucleus and the Process of Fission	10
Figure 0.2	A Nuclear Fuel Assembly	12
Figure 0.3	Diagram of Pressurized Water Reactor and Cooling Tower	13
Figure 4.1	Common Progression of the Different Dimensions of Injustice Associated with the Production of Nuclear Power and Other Unjust Energy Systems	130

TABLES

Table 0.1	Different Paradigms and/or Dimensions of Social Justice	25
Table 0.2	Some of the Antinuclear Power Social Movement Organizations Analyzed in the Present Study	32

Acknowledgments

This project was made possible by many, many people whose love, support, and encouragement carried me throughout the process. I am forever grateful to all those who have made an impact on me throughout my journey.

To begin, Dr. Clarence Lo was and is a tremendous source of encouragement, inspiration, and guidance. At critical points in my intellectual and professional development, Dr. Lo guided me as I struggled to find direction. He recommended key works and helped me explore research options that, without his mentorship, I would have never discovered on my own. His deep understanding of critical theory, political sociology, social movement studies, and other areas has contributed significantly to the present study, as well as my general sociological/theoretical perspective of the world.

Furthermore, this project would not be possible without the help, assistance, and guidance of Dr. Ibitola Pearce, Dr. James "Sandy" Rikoon, and Dr. Rebecca Scott. Dr. Ibitola Pearce helped open my eyes to the breadth and depth of critical race and gender theory, and helped fuel a desire to achieve racial justice as a central goal of my scholarship and teaching. Working with and reading Dr. James "Sandy" Rikoon's work has had a tremendous impact upon the present study, as well as upon my development as an environmental sociologist. Dr. Rikoon's advice and feedback over the years have always helped keep me grounded when my tendency is to sometimes get lost in abstract, theoretical considerations. I am also very grateful to have had the opportunity to get to know and work with Dr. Rebecca Scott. Working as a research assistant for Dr. Scott gave me valuable experience and insight into the research process. Additionally, Dr. Scott's critical theoretical perspective, while very different from my own, has helped sensitize me to different issues of culture and identity involved in political and environmental conflicts. Dr. Scott's mentorship and advice have really helped strengthen this project.

I am also deeply indebted to the rest of the faculty in the Sociology Department at the University of Missouri who helped teach and train me over the course of my graduate studies. I would like to specifically thank Dr. Wayne Brekhus, Dr. John Gallaher, Dr. Jaber Gubrium, Dr. Victoria Johnson, Dr. Amit Prasad, Dr. Joan Hermsen, and Dr. Ed Brent for the opportunity to learn from and work with them. I know that as a professional sociologist I am very much a product of the training I received from these and other faculty. I would also like to thank the MU library staff, especially Sociology Research Librarians Nancy Turner Meyers and Rachael Brekhus—you both helped me so much with finding the materials and resources that I needed.

I am also extremely grateful for the guidance and assistance from my faculty colleagues at Butler University. The support I received from Dr. Carmen Salsbury, Dr. Stuart Glennan, Dr. Carol Reeves, Dr. Julia Angstmann, Dr. Elizabeth Davis, Dr. Elise Edwards, Dr. Kathryn Novak, Dr. Robin Turner, and Dr. Travis Ryan has been invaluable for allowing me to work on this project. I would also like to thank administrative specialists Carla North and Claudia Johnson for all of their help. Lastly, I would like to thank all of the Butler University science, technology, and environmental studies students I have had the pleasure of working with over years. Your thoughts and insights have helped me grow and develop as a scholar and teacher.

I am very appreciative of the anonymous reviewers for their input and feedback on earlier versions of this manuscript. Your comments and suggestions helped strengthen and refine the manuscript and its central arguments. This work is much improved thanks to your contributions. I would also like to thank Courtney Morales and everyone else at Lexington Books for their support, patience, and understanding. Finishing a book manuscript in the context of a global pandemic, while trying to attend to young children, was exceedingly difficult and I am appreciative of all the flexibility everyone at Lexington Books extended to me.

Most importantly, I would have never been able to start (let alone finish) this project without the help, guidance, love, and understanding of my amazing wife and family. My sons Eldon and Sterling have helped me discover a level and intensity of love I did not know was possible before you two came into my life. Your innocence, curiosity, and joy for life are infectious. I hope this book will make you proud, and in some small way make the world you inhabit a better place. My parents, Dennis and Claudia, have both done more for me than I can ever hope to recall, and I love them and am thankful for them more than I can ever express in words. I will never be able to thank you enough or to even express how grateful I am to have you as my parents. Without question, the person most responsible for the completion of this manuscript is my wonderful and loving wife, Mallory Van Gerven. Day-in and day-out over the past fifteen years you have given me strength and love, without which I would have given up long ago. You literally held my hand and rubbed my back as I struggled to find the courage and strength to take the next steps. I love you so much. This is *our* accomplishment. Thank you for everything. To my boys, parents, and my wife Mallory, there is not enough time or space in the world for me to communicate and express to you how much I love you and how thankful I am to have you in my life. What I have written above does not even come close to even beginning to say what I feel. I love you. Thank you.

Finally, to all those who helped along the way but I fail to mention above, my apologies. Do not take it personally and know that your contributions will never be forgotten. And, of course, all errors and mistakes remaining in this work are my responsibility alone.

Introduction

The morning of September 28, 2010, was overcast and cool. Signs of the approaching fall were everywhere in the town of Vernon, Vermont, located on the banks of the southern Connecticut River, three miles north of the Massachusetts state line. Low-hanging clouds gripped the hillsides and gave a sense of forbiddance to the otherwise picturesque New England landscape. As the women approached the gates of the sprawling Vermont Yankee Nuclear Reactor Complex, Frances Crowe knew what they were likely in store for; she had been arrested for her activism for peace and justice many times before. Today she was joined by Hattie Nestel of Athol, Mass.; Ellen Graves of West Springfield, Mass., and Paki Wieland, from Northampton, Mass. The Raging Grannys, as they refer to themselves, were there to block the gates of the reactor complex in an act of protest meant to bring an end to the dirty, dangerous, and expensive production of nuclear power and radioactive waste at the site. In recent years, dozens of people had been arrested for nonviolent civil disobedience at Vermont Yankee and at the Brattleboro offices of Louisiana-based Entergy Corp., which owns the reactor. Many of the protests were organized by the Shelburne Falls, Mass.-based group Citizens Awareness Network (CAN). As the elderly women peacefully used their bodies to obstruct the entrance of the facility, security personnel and local police were called in to remove the protesters. The women were arrested by police and were released the same day with citations to appear in court that December. "There is no such thing as a peaceful, safe, renewable and cost-effective nuclear power facility," Nestel said following her release from jail.

The campus of Loyola University on the north side of Chicago, IL, was largely deserted the first week of June 2010. Most of the university's students had left weeks ago, following the end of the spring semester. In their place slowly arrived over 125 grassroots, antinuclear power activists from 25 states, several American Indian tribes, and 5 countries. We had come to attend the first National Grassroots Summit on Radioactive Waste Policy, organized primarily by Nuclear Information and Resource Services (NIRS), Beyond Nuclear, and Nuclear Energy Information Services (NEIS). The summit provided an opportunity for representatives of community organizations affected by commercial and government nuclear activities to network, organize, workshop, and generally engage in the work of solidarity- and movement-building. The nuclear industry and their allies in government were proudly trumpeting that the United States was on the verge of a "nuclear renaissance"; we had come to Chicago to plan the movement's strategy to prevent a "nuclear relapse." Both the Obama administration and the Department of Energy were actively seeking to increase (by as much as $50 billion) the amount of federal monies available for financing the construction of new nuclear reactors in the United States. The Blue Ribbon Commission on America's Nuclear Future had begun their proceedings, albeit without any representation of community organizations that were impacted by commercial and state nuclear activities. It was clear to those in attendance that the United States was on the verge of a dramatic change of direction with regard to nuclear power and radioactive waste management. We needed to work together to build strategies and tactics to ensure this change was in the direction of increased social and environmental justice. The consensus going into and coming out of the summit was "nuclear power is dirty, dangerous, and expensive." Furthermore, it was agreed that there may in fact not be a scientifically, politically, economically, and/or socially acceptable way to solve the radioactive waste crisis. Thus, the only acceptable course of action going forward was to "STOP MAKING IT!"

On January 25, 2012, 88 national, regional, and local environmental and antinuclear power organizations and more than 5,400 individuals sent a letter to then-Energy Secretary Steven Chu urging him to reject the then-upcoming recommendations from the Blue Ribbon Commission on America's Nuclear Future (BRC) that would encourage the establishment of an "interim" radioactive waste storage dump and begin the transportation of high-level radioactive waste across the United States. The letter was initiated by organizations representing communities around permanently closed reactor sites, such as Citizen Awareness Network of Shelburne Falls, MA, and Citizens Environmental Coalition of Albany, NY. Although the letter was drafted and initiated by seventeen local level, grassroots organizations representing communities living with closed reactors, several national-level antinuclear power organizations utilized their networks, connection, and resources to build stronger support for their claims from dozens of additional organizations and thousands of individuals.

The Commission's draft report cites these closed reactors, which were still storing their waste on their sites, as the reason that an "interim" storage site should be established immediately. As the letter states, such a program runs exactly counter to the interests of these communities,

> The Commission you appointed is claiming that it is acting in the interest of communities such as ours where closed nuclear power reactors are located, when in fact the Commission's recommendations are in opposition to our number one priority: isolation of radioactivity from our environment for as long as it is a hazard. Centralizing waste storage for purposes of expanded waste production or for reprocessing is contrary to this goal, and is not responsible policy.

The groups urged Secretary Chu to "apply the logic that is missing" from the BRC report and prevent the transportation of radioactive waste "until a permanent isolation program is available." The groups also urged the implementation of Hardened On-Site storage of the waste at the reactor sites, as described in a document called *Principles for Safeguarding Nuclear Waste at Reactor Sites*, which has been endorsed by more than 170 organizations. In this and many other cases, the antinuclear power movement in the United States operates on both the national and local levels in their efforts to stop the production of nuclear power and radioactive waste. Furthermore, there is extensive coordination within the movement between these levels as different types of antinuclear organizations work to promote environmentally benign and socially just energy systems.

INTRODUCTION/EMPIRICAL PROBLEMATIC

The grassroots opposition to the continued production of nuclear power and high-level radioactive waste in the United States takes place at every step of the nuclear fuel chain on both the national and local levels. On the *local level*, this opposition takes place in uranium mining communities where local resident groups fight to have the contamination from past mining operations cleaned up and to prevent new uranium mines from opening. Local opposition is also found in communities with uranium fuel-fabrication and enrichment facilities, where local stakeholder groups feel the environmental and human health costs of such facilities outweigh the economic benefits they provide. As seen from the first vignette above, local grassroots opposition is also found in reactor communities where stakeholder groups fight to shut down existing reactors, prevent existing reactors from being relicensed, and prevent new facilities from being licensed and constructed. Finally, local opposition to nuclear energy and radioactive waste is found in (potential) disposal communities, where a broad range of stakeholders work to keep their communities and environments from being turned into the nation's radioactive waste dump.

While there is a great deal of opposition to the continued production of nuclear energy and radioactive waste on the local level, the movement for a carbon-free and nuclear-free energy future in the United States also operates on the *national level*. As demonstrated by the second and third snapshots above, beyond lobbying and other attempts to influence the political process in Washington, DC, national antinuclear social movement organizations (SMOs) in the United States help support and coordinate the activities of local groups around the country. These antinuclear SMOs help educate people about the dangers of nuclear energy and radioactive waste and keep people informed about relevant policy developments and the operations of the nuclear power industry. Additionally, they work to facilitate people's involvement in the political process by distributing action alerts, staging protests, and organizing conference calls and summits for grassroots activists.

As indicated earlier, the politics of nuclear power and high-level radioactive waste management in the United States take place on the local and national levels. Consequently, antinuclear SMOs in the United States have organized and engaged in environmental justice energy activism on both of these levels as well. However, in the United States, policy concerning the development and operation of nuclear power plants and the management of radioactive waste is made at the federal level. In 1942, Enrico Fermi achieved the first controlled nuclear chain reaction at the University of Chicago, and in 1945 the United States used nuclear weapons against the Japanese in WWII. On August 1, 1946, President Eisenhower signed the Atomic Energy Act which guided postwar nuclear weapons and power development under civilian control. This legislation established the Atomic Energy Commission (AEC), which was mandated to both regulate *and* promote the use of the "peaceful atom" for nonmilitary purposes. In 1953, President Eisenhower promotes "Atoms for Peace" to distribute nuclear technology around the world "to serve the peaceful pursuits of mankind," and in 1955 the AEC announced a joint government/industry program to develop nuclear power plants.

However, by the1970s, the oil crisis precipitated by the OPEC oil embargo focused the American public's attention on energy policy for the first time (Perkins, 2017). In this context of heightened concern over energy policy, it was becoming clear the dual mandate of the AEC to both promote and regulate the nascent nuclear power industry created an unsurpassable conflict of interests. Public protest over the handling, storage, and disposal of radioactive waste began in the late 1950s with regard to ocean dumping—a practice that was quickly abandoned due in large part to public outcry (Walker, 2009). Public concern over radioactive waste later skyrocketed after a large and serious leak of radioactive waste was widely reported to have happened at the Hanford, WA, site in 1973. This event lead Ralph Nader and the Union of Concerned Scientists to denounce the AEC's handling of radioactive waste at their own facilities and seriously question their ability to properly regulate commercial facilities (Walker, 2009). Public concern over nuclear power production increased dramatically in the aftermath of the 1979 Three Mile Island accident, and even further following the Chernobyl disaster in 1986.

The Energy Reorganization Act of 1974 abolished the AEC and created the Nuclear Regulatory Commission (NRC). The NRC is an independent government agency responsible for reactor licensing and renewal, the safety and security of operating reactors, and the safe regulation of nuclear materials, including high-level radioactive waste. The NRC's mission is to *regulate* the nation's civilian radioactive materials to ensure the protection of public health, public safety, and the environment. The relationship between the antinuclear movement and the NRC is somewhat complicated, as is often the case with the relationships between environmental/social justice groups and regulatory agencies. On the one hand, the NRC has regularly been accused of falling victim to "regulatory capture," where the industry the agency is supposed to regulate effectively takes control of the regulatory agency itself. It is widely perceived within the antinuclear power movement that the NRC is beholden to the nuclear power industry, and regularly advances the industry's interests over the interests of communities and the environment. On the other hand, antinuclear SMOs often find themselves also having to defend the NRC and the current levels of regulation from efforts to minimize or eliminate both. In the neoliberal era, many instances of governmental regulation of business and industry have come under sustained attack from "free-market" ideologues arguing such regulations are an unjust and inefficient limitation of "freedoms." The NRC has not been an exception. Thus, despite the general level of dissatisfaction felt by the movement toward the NRC, movement actors often find themselves fighting to preserve what minimal regulation and oversight currently exists.

The NRC is targeted primarily by antinuclear groups on the local level concerned with varying issues related to reactor safety, licenses, and license renewal. More specifically, the NRC is often the target of antinuclear campaigns *against* new reactor licenses and license renewals for existing reactors, campaigns *for* better emergency and evacuation preparedness, and *for* better control of the intentional and accidental release of radioactive materials from nuclear power plants. Although national antinuclear SMOs, such as NIRS and Beyond Nuclear, contribute to and assist local level organizations in these campaigns, the campaigns themselves are often driven and led by local organizations representing people and communities most directly impacted by particular nuclear facilities.

The other federal department responsible for the governance of the nuclear industry and nuclear power is the Department of Energy (DOE). The Department of Energy Organization Act of 1977 established the DOE as a cabinet-level department concerned with policies regarding energy, the nation's nuclear weapons program, and the handling of civilian radioactive waste *disposal.* The DOE is headed by the secretary of energy who is appointed by the president. Of great concern for the antinuclear movement is the DOE's responsibility for the handling and storage of the nation's civilian high-level radioactive waste, including the "spent fuel" from commercial nuclear reactors.

Grassroots opposition to nuclear power and radioactive waste production in the United States increased dramatically in the aftermath of the Great East Japan Earthquake of March 11, 2011, and subsequent ongoing radioactive disaster at the Fukushima Diiachi nuclear power plant. Prior to the ongoing Fukushima disaster, no nuclear reactor had been licensed in the United States since the late 1970s and none constructed since the mid-1980s. The Three Mile Island accident in 1979 and the Chernobyl disaster in 1986 soured many Americans' perceptions of nuclear power. According to a *Forbes* magazine cover story in 1985, "the scale of the U.S. nuclear power program's collapse is appalling" and "the failure of the U.S. nuclear power program ranks as the largest managerial disaster in business history, a disaster on a monumental scale" (Cook, 1985).

However, in the first decade of the twenty-first century, it appeared as if this trajectory of nuclear power in the United States might change. In the early 2000s the nuclear industry spent over $300 million on lobbying and P.R. efforts to rebrand nuclear power as the solution to both the problems of "energy independence" and global warming. The industry's efforts were partially realized and rewarded with the passage of the 2005 Energy Policy Act. This Act established the controversial federal "loan guarantee" program for the development and construction of new nuclear reactors. This program was set up to provide taxpayer money in the form of "loan guarantees" to help cover the tremendous up-front costs of obtaining a license for and constructing a nuclear reactor, which runs between $6 and $10 billion. This lead the Nuclear Energy Institute (NEI), the primary trade group for the nuclear power industry, to declare the United States was on the verge of a "nuclear renaissance"; a phrase that was picked up and often repeated by many media outlets. Some nuclear proponents, such as South Carolina Senator Lindsey Graham, claimed the United States would have twenty new reactors by 2020. Between 2005 and 2010, the NRC received twenty-eight applications for "early-site permits" and "combined operating and licensing permits" for new nuclear reactor construction, all of which were subsequently delayed or canceled (NIRS Newsletter, Dec. 2010).

The so-called nuclear renaissance was slow to develop and took some unexpected turns during the Obama administration. In 2010, President Obama cut off virtually all funding for the development and construction of a high-level nuclear waste repository at Yucca Mountain, Nevada, and instructed the DOE to withdraw its license application from the NRC. (Reflective of the relationship between the two agencies/departments, the NRC is tasked with establishing the environmental and human health guidelines for a nuclear waste repository and with issuing the licenses, but it is the DOE that will actually construct and operate the eventual repository.) This was a dramatic change of direction in federal nuclear waste policy, as the Yucca Mountain site has been the only location studied and developed for permanent high-level nuclear waste storage since 1987 at a cost of over $15 billion, according to the Government Accountability Office. This decision also proved to be contentious. Washington state and South Carolina filed a lawsuit against the DOE claiming that the DOE lacks the authority to withdraw their license application. The plaintiffs contended doing so violated the intent of Congress as expressed in the 1987 Nuclear Waste Policy Act Amendments, which designated Yucca Mountain, Nevada, as the site for the nation's first high-level radioactive waste dump. The DOE argued that they did have the authority to withdraw the license application because they are the federal agency responsible for the construction and operation of such facilities.

Along with the termination of the Yucca Mountain project, the Obama administration created the Blue Ribbon Commission on America's Nuclear Future (BRC). The BRC was tasked with answering several key questions regarding the future of radioactive waste disposal in the United States. The BRC was mandated to evaluate possible disposal strategies and issue recommendations in 2011 that were expected to become the basis of new federal radioactive waste policy. The BRC was composed of industry representatives, nuclear engineers and other technical experts, former senators and representatives, policy experts, and labor representatives. Notably absent from the Commission were any representatives from the environmental community, anyone representing impacted communities, or any other opponents of the nuclear industry itself. The Commission's charter emphasized "public participation" and "transparency" would be central in the Commission's proceedings, and both the nuclear industry and their opponents actively participated in and monitored the process.

In addition to the cancelation of the Yucca Mountain project and the creation of the BRC, the Obama administration signaled other shifts in the administration's approach to nuclear energy and radioactive waste disposal. When campaigning in 2008, Obama said he was not a "supporter" of nuclear power, and the only way nuclear power would move forward was if it could compete on the open market, with other energy sources, without government subsidies or taxpayer support. However, in early 2009 President Obama announced he was asking Congress to triple the budget for the "loan guarantee" program (to over $50 billion) that would funnel taxpayer money to help finance the construction of new nuclear reactors. In sum, beginning in 2009 the Obama administration began to dramatically alter the direction of U.S. nuclear energy and radioactive waste policy in ways that seem to be aimed at facilitating a "nuclear renaissance."

This was an extremely important juncture in the history of nuclear power in the United States for both the nuclear power industry and the antinuclear power movement. The decisions and policies enacted in this crucial period would either finally begin-the-end of commercial nuclear power production in the United States or facilitate its dramatic resurgence. However, even with all of this activity favorable for the expansion of the commercial nuclear power at the national level, the "nuclear renaissance" proved to be more a product of proclamation than actualization. This was due in no small part to the efforts of various grassroots campaigns that opposed the nuclear industry and their allies in government on the local and national levels. As it is, no new taxpayer money was in fact allocated for the "loan guarantee" program, despite president Obama's repeated requests from Congress. In 2010 alone, more than 100,000 letters were sent to Congress opposing the expansion of the "loan guarantee" program, along with thousands more phone calls and letters sent to the DOE and White House, and thousands of signatures on petitions (NIRS Newsletter, Dec. 2010). This is all in addition to thousands of people working on various local- and state-level campaigns. Only one project received a "conditional loan guarantee," while a second potential recipient backed away from being awarded a loan guarantee because the financial risks involved were deemed too great. Attempts by local utilities to extend the lifespan of existing nuclear facilities continued to be met with fierce opposition by concerned local stakeholders around the country. Additionally, the routine and accidental releases of radioactive materials from the nation's 96 nuclear reactors (down from 104 at its peak) continue to be a rallying point for activists who demand the closure of these facilities.

And the antinuclear movement was growing on both the national and local levels in these years. For instance, Nuclear Information and Resource Services (NIRS), a central and very active national SMO, reported having a little less than 8,000 supporters in 2008. By December of 2010 they had grown to over 26,000 supporters and 6,000+ more on their Facebook pages (NIRS Newsletter, Dec. 2010). It is clear that as the nuclear industry accelerated its efforts to realize a "nuclear renaissance" in the United States, so too did the movement for a carbon-free and nuclear-free energy future to prevent a "nuclear relapse." In June of 2010, NIRS helped organize the National Grassroots Summit and Forum on Radioactive Waste Policy in Chicago, IL. The summit brought together over one hundred activists from twenty-five states and three countries as a means of encouraging solidarity within the movement, fostering communication and coordination, and building a comprehensive strategy. The Summit/Forum was put together in response to the dissatisfaction felt by many activists regarding the composition of the BRC and the disturbing content of the BRC's first meeting in March 2010. The working consensus that emerged during the Summit/Forum is that there may not be any satisfactory solutions to the problems of safely storing high-level radioactive waste. Therefore, the only sensible course of action is to "stop making more of it!" The movement's message going into and coming out of the summit was "nuclear power is dirty, dangerous, and expensive!" Also, every step of the nuclear fuel chain violates the principles of environmental justice and results in environmental inequalities—be it in mining communities, enrichment/fabrication communities, reactor communities, and/or disposal communities.

The present study critically analyzes the antinuclear power movement (or the movement *for* a carbon-free and nuclear-free energy future) in the United States during these strategically crucial years using an environmental justice framework. The aim is to explore how different conceptualizations/discourses of social and environmental justice were constructed through the claims of antinuclear SMOs facing the prospect of a resurgence of nuclear power. Through a critical analysis of the claims made by the antinuclear power movement in the United States at this time, as well as analyses of the historical/structural conditions these claims were made in response to, general principals of "energy justice" are identified. The identification of general principles of energy justice, similar to the principles of environmental justice, should guide future energy policy to ensure social and environmental justice are maximized. The remainder of this chapter provides a brief overview of how nuclear power is produced, including a discussion of the nuclear fuel chain. Then the relevant literatures from social movement studies, environmental justice, and political philosophies of justice are reviewed. Finally the data sources and research methods utilized in the present study are discussed, along with the significance of the findings.

NUCLEAR POWER 101 AND THE NUCLEAR FUEL CHAIN

So what is nuclear power and where does it come from? To begin to answer these questions it is necessary to start with some basic chemistry. Atoms are constructed like miniature solar systems. At the center of the atom is the nucleus, and orbiting around it are electrons. The nucleus is composed of protons and neutrons, very densely packed together. For example, hydrogen—the lightest element—has 1 proton; uranium—the heaviest natural element—has 92 protons (uranium also usually has 146 neutrons that have no electrical charge, and function to make the nucleus stable). Enormous energy is present in the bonds that hold the nucleus together. This nuclear energy can be released when those bonds are broken, and this energy can be used to produce (generate) electricity. When a nucleus is bombarded with a neutron, it can be split apart, in a process called fission (see figure 0.1). Because uranium atoms are so large, the bonds holding them together are relatively weak, making uranium good for fission. This is specifically the case with uranium-235, a relatively rare isotope that has 143 neutrons, rather than the usual 146.

Uranium is one of the least plentiful minerals on Earth, making up only two parts per million in the Earth's crust (by no means a *renewable* resource). Uranium is found in several geological formations, and is even found in seawater. To be mined as a fuel, however, it must be sufficiently concentrated, making up at least 100 parts per million (0.01 percent) of the rock it is in. As it happens, most of the uranium in the United States is located on Native American lands. Furthermore, many of the largest reserves of uranium globally also happen to be located on indigenous lands (e.g., in Northern Niger and Western Australia). Most of the uranium mines in the United States are on Navajo and Laguna Pueblo Indian lands in New Mexico and Arizona (Grants Mineral Belt), and other locations in the Southwest. Operations at these mines have taken a devastating toll on the native communities near them. Native American miners, most of whom were never informed about the dangers of uranium, were exposed to its particulate and radioactive gases in the mines for decades. They have suffered a large number of lung cancer

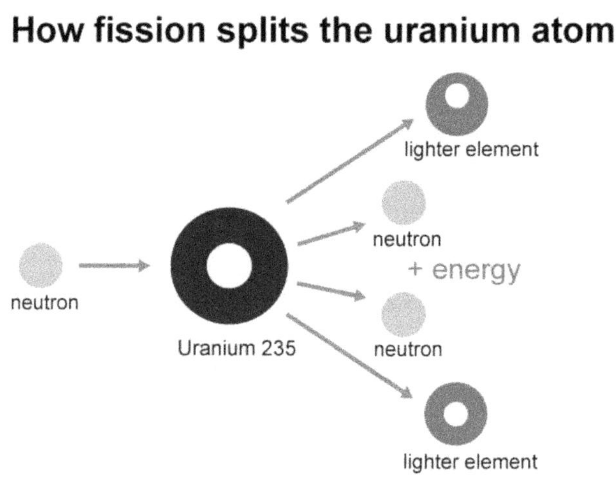

Figure 0.1 Diagram of a Uranium Nucleus and the Process of Fission. Adapted from National Energy Education Development Project.

fatalities, which is a disease that was virtually unknown to the Navajos and Pueblos prior to uranium mining.

The uranium mining process is similar to coal mining, with both open pit and underground mines. It produces similar environmental impacts, with the added hazard that uranium mine tailings are radioactive. Groundwater can be polluted not only from the heavy metals present in mine waste but also from the traces of radioactive uranium still left in the rock. Milling often takes place near mines to reduce the transportation of waste rock. Thus mining debris and mill tailings are often put into unlined storage ponds or piled out in the open air, where they often leach into nearby soil and water. Groundwater that enters the mines, and thus becomes contaminated, is regularly pumped out into nearby lakes and rivers. The situation in the American Southwest was made even worse in the late 1970s when many uranium mines ceased operation because of a steep drop in the price of uranium. Companies across the Southwest abandoned the mines without sealing the tunnels, filling the pits, or removing the large piles of radioactive and toxic tailings. As a result, Native American families have been living in close proximity to these mines and radioactive materials for decades. Tragically, uranium mine tailings were even used to construct roads, homes, buildings, schools, and playgrounds, as part of the infrastructure development that was promised to the Navajos as compensation for the uranium extraction on their lands. Many serious health effects, beyond lung cancers, have been extensively documented.

As found in nature, uranium is more than 99 percent U-238; unfortunately, U-235, a rare isotope, is what is used in power plants. So, once it is mined and milled, the uranium ore is sent to a processing plant to be concentrated into a useful fuel. There are sixteen processing plants in the United States, although eight are inactive. Most uranium concentrate is made by leaching the uranium from the ore with acids, which creates massive amounts of vary dangerous waste. When finished, the uranium ore is turned into U-3O8, the fuel form of uranium, and formed into small pellets. The pellets are packed into 12-foot long rods, called fuel rods. The rods are bundled together into fuel assemblies, ready to be used in the core of a reactor (see figure 0.2).

In the reactor of nuclear power plants, neutrons collide with uranium-235 atoms, splitting them. This split releases neutrons from the uranium that in turn collide with other atoms, causing a chain reaction. This chain reaction is controlled with "control rods" that absorb some of the freed neutrons. Without these control rods, the chain reaction would go "critical" and explode as in a nuclear bomb. In pressurized water reactors, the fission releases energy that heats pressurized water to about 520°F in the core of the plant. This hot water is then used to make steam to spin turbines that are connected to generators that produce electricity. There are currently 96 commercial nuclear reactors in operation in the United States (down from 104 in January 2013).

Figure 0.2 A Nuclear Fuel Assembly. *Source:* Alternative Energies and Atomic Energy Commission, France.

Commercial nuclear power plants range in size from about 60 megawatts for the first generation of plants in the early 1960s, to over 1,000 megawatts, and many plants contain more than one reactor. In the United States, two-thirds of the reactors are pressurized water reactors (PWR) and the rest are boiling water reactors (BWR). In a BWR, the water is allowed to boil into steam and is then sent through a turbine to produce electricity. In PWR, the core water is held under pressure and not allowed to boil. The heat is transferred to water outside the core with a heat exchanger (also called a steam generator) and the outside water boils into steam and drives a turbine. In PWR, the water that is boiled is separate from the fission process, and so does not become radioactive (see figure 0.3).

After the steam is used to power the turbine, it is cooled off to make it condense back into water. All plants use tremendous amounts of water (in some cases millions of gallons per minute) from rivers, lakes, or the ocean to cool the steam. Some do this with the iconic hourglass-shaped cooling towers, while others dump the hot water back into nearby bodies of water. The plants that do not use cooling towers are almost all located near the ocean, which is used as a big heat sink. For example, Ameren UE's nuclear power reactor in Callaway County, Missouri, withdraws huge amounts of water from

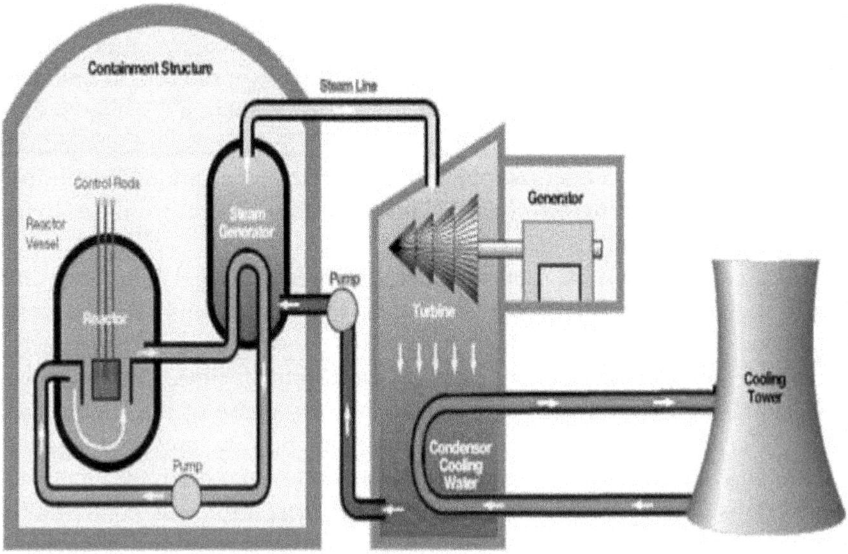

Figure 0.3 Diagram of Pressurized Water Reactor and Cooling Tower. *Source:* United States Nuclear Regulatory Commission.

the Missouri River, evaporates most of it in the cooling tower, and returns the remainder—hotter, but not nearly as hot as it would be if it was used in a once-through loop in boiling water reactor. The hourglass-shaped cooling towers are the familiar landmark of many nuclear plants. For every unit of electricity produced by a nuclear power plant, about two units of waste heat are rejected to the environment.

About 95 percent of all radioactivity created in the United States, from all sources including nuclear weapons production, is contained in the irradiated fuel—or high-level radioactive waste—from commercial nuclear power reactors. Yet even so-called low-level nuclear waste can contain lethally radioactive and long-lived elements, such as Plutonium-239, Strontium-90, and many others. Radioactive waste is, fundamentally, the lethal byproduct of the nuclear age. The storage and transport of high-level radioactive waste (HLRW) is one of the most controversial dimensions of commercial nuclear power production. No nation has yet solved the problem of what to do with this material, which must be shielded from all living things and the environment for millennia. Highly radioactive wastes include solid irradiated nuclear fuel assemblies (euphemistically called "spent" or "used" by the industry that creates them) and liquid HLRWs resulting from "reprocessing." Irradiated nuclear fuel rods discharged from commercial nuclear power plants are highly radioactive, a million times more so than when they were first loaded into a reactor core as "fresh" fuel. If unshielded, irradiated nuclear fuel just removed from a reactor core could deliver a lethal dose of radiation to a person standing three feet away in just seconds. Even after decades of radioactive decay, a few minutes of unshielded exposure could deliver a lethal dose. Certain radioactive elements (such as plutonium-239) in "spent" fuel will remain hazardous to humans and other living beings for hundreds of thousands of years. Other radioisotopes will remain hazardous for millions of years. Thus, these wastes must be shielded and isolated from the living environment for hundreds of millennia.

Recent estimates put the total volume of commercial HLRW at more than 70,000 metric tons—10 percent more than the storage limit for the now-abandoned Yucca Mountain Repository. This waste is being "temporarily" stored in cooling tanks and dry-casks at the reactor facilities where it was produced. Radioactive waste is generated at every phase of the nuclear fuel chain. Whether high-level or so-called low-level, no safe, proven, long-term solution has yet been found to manage these wastes. In other words, radioactive waste is a problem because, to varying degrees, it remains highly radioactive for very long periods of time. Radiation causes serious, even fatal, human health and environmental problems, and there is no known technically and/or politically acceptable way to keep it separated from the environment, and consequently human populations. The term "disposal" simply cannot be applied to materials that will remain hazardous virtually forever. There is no place on Earth where one can confidently predict that radioactive waste could remain safely isolated from the environment for hundreds of thousands to millions of years.

Over the past thirty to forty years, international scientific consensus has more or less settled on "deep-geological disposal" as the best *available* solution to dealing with radioactive waste. In 1987, Yucca Mountain, Nevada, was singled out as the only site for further study in the United States for the development of a permanent high-level radioactive waste repository (HLRWR). Many have argued that the selection of Yucca Mountain was not due to the technical and/or geologic suitability of the site for an HLRWR, but rather the political vulnerability of the State of Nevada in the late 1980s. For example, Yucca is located in a major earthquake zone. Dozens of fault lines crisscross the area, with two directly intersecting the proposed dump-site. Many hundreds of quakes have struck near Yucca Mountain in recent decades, damaging DOE facilities and derailing trains that could one day be used to haul nuclear waste. With the cancelation of the Yucca Mountain Project in 2009 and the inception of the BRC in 2010, the United States was at a crossroads with regard to radioactive waste policy.

Radioactive contamination and social and environmental injustices accumulate at every step of the nuclear fuel chain. This is especially true at the front-end and the back-end of the nuclear fuel chain, in uranium mining communities, and (eventually) radioactive waste storage communities. The antinuclear power movement seeks to put an end to this production chain that costs communities, utility ratepayers, and taxpaying citizens billions of dollars, and creates significant, negative human health and environmental impacts. Antinuclear power organizations seek to halt the "dirty, dangerous, and expensive" production of nuclear power by agitating for energy justice, which in part seeks to include the full consideration of the societal and environmental impacts of energy production into decisions about energy policy. In order to critically analyze these efforts, it is necessary to review the relevant scholarly literatures in the areas of social movement studies, environmental justice, and political philosophy.

SOCIAL MOVEMENT STUDIES, ENVIRONMENTAL JUSTICE, AND MORAL PHILOSOPHY

Social movement theory and research has changed and developed a great deal over the past half-century as new concepts and theories have been introduced in order to explain and understand the emergence of seemingly new kinds of social protest. In the early part of the twentieth century, many scholars viewed protest and social movement activity as irrational and irregular. The "myth of the madding crowd" and the imagery of "group mind" dominated "mass society" theories where protesters were seen as somehow deficient or insufficiently attached to social institutions, and protest itself was seen as "moments of madness" (Jasper, 1997). Then in 1977, McCarthy and Zald published the enormously influential article "Resource Mobilization and Social Movements: A Partial Theory" which established social movement activity was a normal part of politics as opposed to an aberration within it. Furthermore, they showed social movement activists were "normal" people pursuing reasonable goals, and economic resources helped determine what social movements were able to accomplish. "Resource mobilization theory," as it has come to be called, has spawned an enormous volume of research that together forms a sort of "how-to" manual for social movement activists and organizations by focusing on how activists finance their activities, what tactics are more successful than others, and so on.

In general, these studies tended to be very economistic in outlook, and focused on how individuals, groups, and organizations calculate costs and benefits (Zald and McCarthy, 1987). By focusing on the resources needed to maintain large, formal organizations the resource mobilization tradition has been amazingly successful at developing and testing specific hypothesis and generating new concepts for the study of SMOs (Zald and McCarthy, 1987; Piven and Cloward, 1979). For example, Walsh (1981) uses the rapid growth of SMOs around Three Mile Island after the 1979 nuclear accident as data for assessing the strength of resource mobilization theory. He finds evidence for the significance of grievances for precipitating and sustaining social movement activity. He concludes grievances, existing structures, and mobilizing processes should all be considered variables (rather than constants) in seeking to develop better social movement theory.

However, if early social movement theories recognized social movement activity as that of angry, irrational mobs, some versions of resource mobilization theory error to the other extreme by presenting an image of social movement activists and organizations as purely self-interested, rational, and calculating entities whose behaviors can be explained and predicted formulaically. This is where the "political process" models of Tarrow (1994) and McAdam (1982) enter conceptually by recognizing the importance of economic resources, but also recognizing the importance of broader historical, political, economic, and social contexts of mobilization activities. More specifically, political process models demonstrated how important "political opportunity structures" were in determining the success or failures of various social movements. McAdam (1982) identifies three factors that are said to significantly impact a social movement's chances of achieving its goals; changing opportunities in the political environment; the existing level of organization in the community; and "cognitive liberation" which he describes as the people's (individual and collective) assessment of their chances for success. McAdam focuses primarily on the opportunities provided by the state with regard to the occasional slackening of repressive practices, which otherwise undermine the efforts of social movement activity. Tarrow (1994) defines "political opportunity structures" more broadly as

> consistent—but not necessarily formal or permanent—dimensions of the political environment that provide incentives for people to undertake collective action by affecting their expectations for success or failure. Theorists of political opportunity structure emphasize the mobilization of resources external to the group. (pp. 85)

Here we can clearly see the difference in focus between the "resource mobilization" theorists and the "political opportunity" theorists, regarding resources and strategy.

Related to the political process model, especially McAdam's discussion of "cognitive liberation," is a body of social movements literature concerned with the issues of "framing" and "frame analysis." Framing refers to the process by which individuals, groups, and organizations identify, interpret, and express social and political preferences and grievances. Framing can also be viewed as a way of organizing interpretations, which guides the ways beliefs and meanings are put together and expressed by movement activists (Taylor, 2000). Most relevant to social movements are "'collective action frames' [which] are emergent, action-oriented sets of beliefs and meanings developed to inspire and legitimate social movement activities and campaigns designed to attract public support" (Taylor, 2000, pp. 511; also see Snow and Benford, 1992). Frame analysis in social movement studies did a great deal to correct for the overly structural focus of both the resource mobilization theories and political process theories, by focusing on issues of collective identity, culture, and symbolically mediated understandings. For instance, the concept of "frame alignment" refers to the process of linking the individual's interpretive framework with that of the social movements, and is absolutely necessary for social movement recruitment and mobilization (Taylor, 2000; Snow and Benford, 1986). According to Capek (1993) and Taylor (2000) "environmental justice," for instance, has emerged as a "master frame" used to mobilize activists who want to link racism, injustice, and environmental concerns under one conceptual frame. Master frames are styles of punctuation, attribution, articulation, and/or ideological frameworks that provide the interpretive medium through which activists identify problems, assign blame, and propose solutions (Taylor, 2000; Snow and Benford, 1992).

As Taylor (2000) argues, it is difficult to examine mass movements that occur over a period of time with a single theoretical lens, because almost by necessity different theories are sensitive to different dynamics involved. Therefore, a number of social movement scholars have begun to synthesize the three theoretical approaches discussed earlier (McAdam et al., 1996). This multifocal theoretical approach is what guides the current analysis of the claims made by SMOs working for a nuclear-free future. More specifically, the present study will contribute to the literature on social movements by analyzing how material *and* cultural resources are acquired and mobilized by SMOs on the local and national levels in their fights to oppose the continued production of nuclear power and radioactive waste. It addresses questions of how resources are conceptualized and utilized in the course of social movement activity, and to what effect. Additionally, following Pellow (2007) a "political-economic opportunity structure" approach is utilized that recognizes political opportunities are not just provided/produced by the state, but also by corporations and other economic actors who play an increasingly formative role in shaping SMO's perceptions of their chances for success or failure. This approach is especially valuable for understanding how SMOs interpreted, constructed, and responded to the Fukushima disaster as an opportunity to advance their goals of ending nuclear power.

Most significantly, the present study presents a thorough analysis of how the problems of nuclear energy and radioactive waste are framed by SMOs on the national and local levels. Broadly, the movement frames the problems with nuclear energy as being "dirty, dangerous, and expensive." The analysis explores how these frames are developed and deployed in the claims made by SMOs on different analytic levels to better understand the social and environmental injustices associated with the production of nuclear power. This in turn will allow for the identification of general features of energy justice, which should guide future energy policy and decision-making. For instance, the "nuclear power is expensive" frame dominates the movement's claims on the national level when addressing issues of reactor financing and the federal "loan guarantee" program for new reactor construction. The movement's claims utilizing this frame primarily draw from and reconstruct the social justice paradigm of distribution (discussed in the following paragraphs). These claims highlight the distributive injustice of public financing for the construction of privately owned facilities, which effectively socialize the risks/costs while privatizing the benefits/profits. This indicates that the development of environmental and socially just energy systems, or energy justice, requires economic and financing procedures for new infrastructure development that break from this model of publically financing the private accumulation of capital.

Related to issues of framing is the fact that many antinuclear energy SMOs also frame the problem of nuclear energy production as an example of environmental racism, and work to incorporate principles of environmental justice into their claims-making and the decision-making processes themselves. This is due to the fact that many of the burdens associated with nuclear energy production (contamination from uranium mining and uranium enrichment/fabrication facilities, the routine and accidental release of radioactivity from nuclear reactors, the burdens of radioactive waste disposal, etc.) are concentrated in communities of color and other low-income areas (Chen, 2011). While the terms "environmental justice" and "environmental racism" are relatively new, the dynamics of environmental inequality and injustice, which the environmental justice movement (EJM) primarily seeks to overcome, are much older. From the movement's very humble beginnings in places like Warren County, NC, and others like it in the early 1980s, the movement has developed into a powerful social force for organizing communities of color and poor and working-class white communities concerned with the increasing environmental pollution and degradation in the places where they live, work, pray, play, and go to school. The EJM has to a remarkable extent managed to redefine environmentalism as intricately connected with the *social* needs of all people, which "brings the environment home," so to speak. As Capek (1993) observes,

> representing more economically marginal actors than do traditional "mainstream" environmental organizations, these grass roots groups are more likely

> to frame their demands in terms of social justice and to challenge stratification based on race, class, gender, and the distribution of power. Accordingly, the concept of "environmental justice" has emerged as a prominent part of the national dialogue over citizen empowerment and the environment. (pp. 5)

In the late 1970s and early 1980s, local environmental justice groups began forming in predominantly white working-class communities (such as the Love Canal area of Niagara Falls, NY) and communities of color (such as Warren County, NC) that were experiencing potentially lethal environmental degradation, primarily from toxic pollution. As Bullard (1990) argues, "in many instances, grassroots leaders emerged from groups of concerned citizens (many of them women) who see their families, homes, and communities threatened by some type of polluting industry or government policy" (pp. 8). The movement grew a great deal during the 1980s as local groups and activists began collaborating with one another at regional and national gatherings (culminating in the 1991 First National People of Color Environmental Leadership Summit) where they developed and shared strategies, tactics, and ideas.

> By the early 1990s the term "environmental racism" had caught fire in social movement, scholarly, policy, and media networks and the call for environmental justice had galvanized one of the most exciting and hopeful social causes of the twentieth century. Environmental justice advocates were regularly engaging polluting corporations, regulators, the courts and elected officials in city councils, in state legislatures, in Congress, and in the White House. (Pellow and Brulle, 2005, pp. 8)

Without doubt, the EJM has significantly altered the directions that environmental research, policy, and activism have taken in the United States. The movement has succeeded in capturing the attention of many high-ranking government officials, as evidenced by President Clinton signing Executive Order 12898 that mandated all federal agencies ensure environmental justice in their operations. Additionally, despite the many problems and power imbalances associated with high-level advisory committees such as the National Environmental Justice Advisory Council (NEJAC), the existence of such committees, at the very least, reflects recognition of the issue in the political process. Having said this, it is on the local level that the EJM has secured its most concrete and clearest victories, and consequently this is the level where the movement seems to have the most influence and power. The literature is filled with many examples of community activists successfully shutting down noxious incinerators, landfills, and other waste storage facilities in their communities (Bullard, 1993, 2005). Additionally, there are a number of cases where communities successfully prevented polluting facilities from being built or expanding (Bullard, 2005) and successfully secured federal buy-outs and/or relocations from proven contaminated communities (Capek, 1993; Blum, 2008). The environmental justice frame has been

> flexible enough to allow Blacks, Native Americans, Latinos, Asians, and Whites concerned with social justice issues to fashion campaigns and develop policies around environment and inequality. Because of its concern with race, class, and gender inequality, the frame was attractive to a wide range of people, yet at the same time, it kept the potency and focus on the environmental inequality of marginalized people. (Taylor, 2000, pp. 562).

This multidimensional flexibility and wide-range appeal have allowed antinuclear SMOs to adapt the environmental justice frame in creative new ways to address the adverse effects of nuclear energy on people of color and other low-income communities. Furthermore, the current analysis of these adaptations and the uses of discourses of justice in the claims made by the antinuclear movement will help identify general attributes of socially and environmentally just energy choices.

Current conflicts over nuclear energy policy have important social dimensions when viewed through the frameworks of environmental racism and the EJM. This framework links local and community-based activism with more broad-based energy activism on national and international scales (Sze, 2007). Energy (and energy policy) is typically regarded as being a technological issue that is the exclusive domain of technocratic and scientific experts. However, energy policy is also given social and political meanings as it is contested and culturally shaped (Nye, 1998; Tatum, 2000). Energy is racialized through the political processes surrounding the practices of regulation, deregulation, and distribution (Sze, 2005). These racialized dimensions of energy policy can most clearly be seen through the claims made by opponents of nuclear energy regarding the negative consequences of nuclear energy development. Environmental justice energy activism is a manifestation of particular interpretive response of activists and SMOs to the implementation of energy systems that are particularly destructive toward communities of color and low-income communities (Sze, 2005, 2007). Through their claims-making, activists and SMOs transform what is normally considered a purely technological issue into a political and moral subject that is open to debate (Sze, 2007). The present analysis of how these types of claims are raised by SMOs on different analytic levels provides new insight into the process of racial environmental inequality formation (Pellow, 2007), collective interpretations, responses and actions on different scales, and how different conceptualizations of "justice" inform the claims made.

In this way, through the use of the environmental justice frame and claims that draw from different dimensions of social/environmental justice, activists use the issues of nuclear energy to address other larger problems of democracy and justice. This allows for the analysis of the antinuclear movement's claims to be used to begin the process of establishing what principles of "energy justice" would look like in practice, and how the United States could move toward such arrangements. More specifically (and discussed in greater detail in the following paragraphs), the analysis of the antinuclear power movement's claims on the national level regarding the federal "loan guarantee" program for new reactor financing reveals the distributive, recognitional, and representative injustices involved in this program. Second, the analysis of the antinuclear power movement's claims on the national level regarding the management of the nation's 70,000+ tons of high-level radioactive highlights the deeply rooted representative, recognitional, and (eventual) distributive injustices that have plagued the nuclear waste management in the United States since the beginning and continue to hamper the program's progress. Third, the analysis of the antinuclear power movement's claims on the local level regarding "construction work in progress charges" (CWIP) for the financing of new reactor construction and issues of reactor safety reveals similar and different representative, recognitional, and distributive injustices that emerge from local/community standpoints with regard to nuclear projects and facilities. Finally, the identification of these multidimensional injustices associated with different aspects of nuclear power production facilitates the proposal of general features (or principles) of "energy justice." The principles of energy justice derived from these analyses, similar to the principles of environmental justice, should be used to guide future energy policy in ways that ensure the protection of social interests, public health and the environment remain central in future decisions about energy.

In order to carry out such an analysis, however, it is necessary to go beyond the traditional, empirical foci of social movement studies discussed earlier. As Somers (2008) argues in *Genealogies of Citizenship: Markets, Statelessness, and the Right to Have Rights* there has been a disciplinary division of labor between sociologists who write about people who fight *for* rights and justice and political philosophers who write about the *meaning* of those rights and conceptions of justice. While too often sociologists respect this intellectual turf divide, social movements and their participants most certainly do not. Social movement actors and organizations almost always express their motivations and goals in terms of rights, justice, and entitlement. Although concern with resource mobilization, political-economic opportunity structures, and framing are essential for understanding social movement activity, "they should not crowd out attention to the more normatively driven powers of justice and rights. In a pluralist knowledge culture, the empirical and the normative are mutually interdependent" (Somers, 2008, pp. xiii). (In)justice, when viewed through the claims made by social movement actors and organizations, is both a normative and an empirical concept. As much as social justice is an empirical arrangement between groups and individuals, it is also normatively perceived as an a priori good, and this normative perception has causal powers (Somers, 2008).

But what exactly are the empirical and normative conceptualizations of justice that are constructed in the course of social movement activity? Nancy Fraser (1997, 2000, 2001) explores "folk paradigms" of two interrelated conceptions of justice; those concerning issues of "distribution" and those concerning issues of "recognition." These efforts culminated in Nancy Fraser's and Axel Honneth's (2003) work, *Redistribution or Recognition? A Political-Philosophical Exchange*, where Fraser's early theorizing found its fullest expression. According to Fraser (2003) "recognition" has become a keyword of our time. This category of Hegelian philosophy has been resuscitated by political theorists in their efforts to conceptualize today's struggles over identity and difference. On the other hand, "redistribution" was central to social struggles and philosophies of egalitarian liberalism in the Fordist era. This paradigm of distributive justice seemed well suited to analyzing the claims of labor and the poor in that period. Conflicts turned primarily on resources and were disputed in distributive terms, with questions of difference usually being relegated to the sidelines (Fraser, 2003). The terms "recognition" and "redistribution" refer to families of claims made by political actors and social movements. These folk paradigms are often presupposed by social movement and political actors and are comprised of sets of liked assumptions about the causes of and remedies for injustice.

According to Fraser,

> Once the hegemonic grammar of political contestation, the language of redistribution is less salient today . . . thanks to a sustained neoliberal rhetorical assault on egalitarianism, the absences of any credible model of "feasible socialism," and widespread doubts about the viability of state-Keynesian social democracy in the face of accelerated economic globalization, they [movements who until recently boldly demanded an equitable share of resources] have ceded pride of place to movements focused chiefly on recognition. (Fraser, 2003, pp. 21)

This creates what Fraser calls the "problem of displacement" wherein the shift from redistribution to recognition struggles is occurring in the context of acceleration of economic globalization and resulting economic inequities. "Thus, cultural conflicts have achieved paradigmatic status at precisely the moment when an aggressively expanding capitalism is exacerbating economic inequality . . . recognition struggles are serving less to supplement, complicate, and enrich redistribution struggles than to marginalize, eclipse, and displace them" (Fraser, 2003, pp. 22).

This problem is further exasperated by what Fraser calls "Hegelian identity politics." The identity model, which is the most common theoretical approach used to study recognition struggles, begins with the Hegelian idea that identity is constructed dialogically through a process of mutual recognition. Proponents of the identity model transpose the Hegelian recognition schema onto the cultural and political terrain. "They contend that to belong to a group that is devalued by the dominant culture of one's society is to be misrecognized, hence to sustain damage to one's individual and collective identity" (Fraser, 2003, pp. 23–24). According to Fraser, the main problem with this is it abstracts misrecognition from its institutional matrix, which obscures its entwinement with distributive justice. By concealing the links between recognition and distribution, misrecognition is stripped of its social-structural (empirical) underpinnings. For example, this approach misses the links, institutionalized in labor markets, between androcentric norms that devalue activities coded as feminine, on the one hand, and female workers' low wages, on the other. Similarly, this approach misses the links, institutionalized in the social welfare system, between heterosexist norms that delegitimate homosexuality, on the one hand, and the denial of resources and benefits to gay and lesbian couples, on the other (Fraser, 2003). As an alternative way of thinking about the role of recognition in democratic practices, Fraser proposes what she calls the "status model of misrecognition." Whereas "identity politics" sees misrecognition as stemming from "free-floating discourse" that fosters misrecognition, she argues for an approach that sees misrecognition as stemming from *institutionalized* patterns of cultural value.

> My proposal is to treat recognition as a question of *social status* . . . what requires recognition is not group specific identity but rather that status of individual group members as full partners in social interaction. Misrecognition . . . does not mean the depreciation or deformation of group identity. Rather, it means social *subordination* in the sense of being prevented from participating as a peer in social life. (Fraser, 2003, pp. 27 emphasis added)

This is a model that seeks institutional remedies for what it sees as institutional harms, which are understood to be based on concrete social relations.

This theoretical model is correct as far as it goes. However, following Fraser's (2009) book *Scales of Justice: Reimagining Political Space in a Globalizing World*, scholars have begun to realize seeing issues of distribution and recognition as the *sole* dimensions of justice only make sense if the nation-state and associated predetermined citizenries are assumed to be the appropriate unites of analysis regarding questions and practices of democracy and social justice. Once the "frame" of justice becomes the subject of contestation, as it increasingly has in transnational movements for environmental justice and human rights (Blau and Moncada, 2006), the third dimension of justice and democracy, what Fraser calls the "political dimension," comes into view (Fraser, 2009). The political dimension of justice deals with politics in its

> constitutive sense, which concerns the scope of the state's jurisdiction and the decision rules by which it structures contestation. The political in this sense furnishes the stage on which struggles over distribution and recognition are played out. Establishing criteria of social belonging, and thus determining who counts as a member, the political dimension of justice specifies the reach of those other dimensions. (Fraser, 2009, pp. 17)

This means who is included in, and who is excluded from, the population of those who are entitled to a just distribution and reciprocal recognition is determined politically, and cannot be understood by considering issues of redistribution and recognition alone.

As is indicated in the aforementioned quote, this dimension of justice is focused on issues of "membership" and "procedure" and is primarily concerned with "representation"—as a matter of social belonging and as procedures that structure public processes of contestation. At both levels, according to Fraser (2009), it is possible to question whether the "relations of representation are just." For instance, do the boundaries of a political community wrongly exclude some who are entitled to representation? And, do the community's decision-making rules accord equal voice and fair representation in public deliberations to all members? According to Fraser (2009), these issues are conceptually distinct from issues of redistribution and recognition, while being heavily interrelated with them in practice, which requires that we adopt a three-dimensional view of justice. Because "representation" is the defining issue of the political dimension of justice, the characteristic political injustice is "misrepresentation." "Misrepresentation occurs when political boundaries and/or decision rules function wrongly to deny some people the possibility of participating on par with others in social interactions" (Fraser, 2009, pp. 18). What she calls "ordinary-political misrepresentation" occurs when "decision rules function wrongly," as for instance when gender-blind rules, in combination with gender-based maldistribution and misrecognition, function to deny women parity of political participation (Hawksworth, 2006, pp. 220–244).

> Less obvious, perhaps, is a second level of misrepresentation, which concerns the boundary-setting aspect of the political. Here the injustice arises when the community's boundaries are drawn in such a way as to wrongly exclude some

> people from the chance to participate *at all* in its authorized contests over justice. (Fraser, 2009, pp. 19 emphasis added)

Frame setting, according to Fraser, is among the most consequential of all political decisions, and she calls this form of injustice "misframing." The consequence of misframing is a kind of "meta-injustice" where the victim is denied the chance to press first-order claims (for recognition and distribution) in a given political community. The injustice remains even if the victim is included as a subject of justice in another community—as long as the effect of the political division is to put some relevant aspects of injustice beyond their reach. Most serious of all is the case where someone is excluded from membership in *any* political community, which is according to Fraser (2009) the denial of what Arendt (1973) called "the right to have rights." Table 0.1 summarizes the basic features of the three dimensions/paradigms of social justice identified in Fraser's three-dimensional model.

In order to develop such a theory, it is necessary to return to an analysis of social movement activity, because it is here that the normative and empirical foundations of a robust theory of justice are interpreted, constructed, disputed, and disseminated. David Schlosberg's (2007) *Defining Environmental Justice: Theories, Movements, and Nature* seeks to understand how movements for environmental justice define the "justice" they seek. Schlosberg asks, are the discourses and conceptualizations of justice that emerge from the EJM different from the distributional conceptions of justice that have dominated moral philosophy since Rawls. It's not that distributive theories

Table 0.1 Different Paradigms and/or Dimensions of Social Justice

Paradigms of Social Justice	*Corresponding Injustice*	*Primary Social Dimension/Sphere*	*General Logic of Claims*
Distribution	Maldistribution	Economic	Removal of unjust difference; economic, environmental, etc.
Recognition	Misrecognition	Cultural	Positive, institutional revaluation of just difference
Representation	"Ordinary" Misrepresentation and Misframing	Political	Sets the framework for claims for recognition and distribution

don't apply to the EJM. The problem is developments in political theory (i.e., Fraser 2003, 2009) have not been applied to the EJM. Through the application of Fraser's theoretical framework (as well as other theories of moral and political philosophy), Schlosberg concludes the EJM does not have a single definition of justice, rather, it articulates many. Justice is about distribution, but it's also about individual and community recognition, participation, and functioning. Groups emphasize different notions of justice on different issues in different contexts, so there is a flexible, heterogeneous, and plural discourse of justice. More specifically,

> movements use a wide range of conceptions of justice, and we can find arguments in those movements for distribution, recognition, participation, and capabilities. The EJM supplies ample evidence that all of these conceptions of justice are used in *practice*, and that, in fact, a comprehensive understanding of the way that movements define the "justice" of "environmental justice" must include all of these discourses. (Schlosberg, 2007, pp. 5)

The present study seeks to bring together and advance the insights of Sze (2005, 2007), Somers (2008), Fraser (2005, 2009), and Schlosberg (2007) through an analysis of the claims made by the antinuclear power movement in the United States. This study extends Schlosberg's (2007) analysis of how discourses and/or conceptualizations of justice underlie, and are constructed through, the political claims-making of the EJM to those of the antinuclear power movement in the United States. In this way, the present study conceptualizes the antinuclear power movement as an example of what Sze (2005, 2007) calls "environmental justice energy activism," which is a distinct and emergent form of environmental justice activism that links traditional local level, grassroots environmental justice activism to energy activism on broader (national) levels. Examining the claims made by environmental justice energy activists, including those of the antinuclear power movement, reveals the multiple dimensions of the social and environmental injustices associated with dirty, dangerous, and expensive energy systems. This is because, as Somers (2008, pp. xiii) has argued, "In a pluralist knowledge culture, the empirical and the normative are mutually interdependent." This means what is normatively held to be just is not determined in isolation from historical and existing social arrangements, social hierarchies, and differentials in social power. In other words, what *is* just is not determined by beginning with abstract principles of justice derived in isolation from extant social formations, as is characteristic of liberal, Western moral philosophy following from Rawls. Rather, what *is* normatively just becomes determined through the course of social interaction, communication, and conflict in empirical, historical, and political-economic contexts.

In his review of moral-philosophical theories of justice, Schlosberg (2007) notes that of all the theorists he reviewed only Fraser (2003, 2005) began her theorizing with an analysis of real-world struggles against existing injustices, which enabled her to develop a more broad and robust theory of justice. Thus, Fraser's work establishes the utility of a methodological approach that uses analyses of real-world claims for justice (or against injustice) as the foundations for the development of better and more inclusive theories of justice. This methodological approach seems to be somewhat dialectical in nature, in that just and unjust social arrangements are not seen as conceptually separate entities, the nature of each being established by philosophical reflections on abstract principles. Rather, just and unjust social arrangements represent different poles within a single totality, and what is held to be normatively just is determined in relation to historical and existing empirical realities. Therefore, the analysis of the claims made by the antinuclear power movement in the United States can not only help identify the various dimensions of the social and environmental injustices associated with current energy systems but also, and more significantly, can help begin to establish general principles of energy justice that seeks to maximize social and environmental interests with regard to energy.

The primary aims of the present study and the central research questions follow from the above. More specifically, these include:

- How are material and cultural resources utilized and conceptualized by the antinuclear power movement? How do the multiple and complementary framings of the issue act as discursive resources?
- How do the claims made by the antinuclear power movement in the United States draw from and rearticulate the social justice paradigms of distribution, recognition, and representation on the national and local levels?
- How do claims regarding (un)just distribution, recognition, and representation (with regard to the production of nuclear power and radioactive waste) reveal the significance of the multidimensional injustices associated with this form of energy production?
- How can the identification of these multidimensional injustices contribute to the work of both correcting existing injustices and, more importantly, developing socially and environmentally just alternatives in the future?

Therefore, the present study and analysis not only works to highlight the injustices of current energy systems but also to identify principles of energy justice that must guide future policy and decision-making in more socially and environmentally just directions.

The analysis that follows first analyzes the antinuclear power movement's claims on the national level, with regard to the primary issues of new reactor construction through "federal loan guarantees" and the program for managing the nation's 70,000+ tons of HLRW. More specifically, the analysis centers on the movement's claims regarding the maldistribution, misrecognition, and misrepresentation involved with the public financing of new nuclear reactors and facilities. This analysis allows for the identification of multidimensional injustices associated with the "loan guarantee" program, and in the concluding chapter, the distillation of general principles of energy justice regarding the financing and implementation of energy systems. Additionally, the analysis focuses on the antinuclear movement's claims with reference to the misrepresentation, misrecognition, and ultimately maldistribution involved with past, current, and future efforts to manage the nation's HLRW. This analysis helps highlight a number of the representative, recognitional, and distributive injustices associated with past, current, and future attempts to manage nuclear waste, which have severely hampered progress from the very beginning. As before, this analysis also enables the identification of several additional features of energy justice concerning the minimization of hazardous material production and the concentration/distribution of benefits/burdens associated with energy systems.

Following this, the analysis turns to the claims made by the antinuclear power movement on the local level, in order to present as complete a picture of antinuclear activism in the United States as possible. In order to maintain consistency between the analyses of the two levels of the movement in the United States, the analysis again focuses on issues of new reactor financing and issues having to do with the more proximate dangers posed by nuclear facilities. More specifically, the analysis turns to the ways local level antinuclear organizations utilize discourses of distribution, recognition, and representation to highlight the multidimensional injustices associated with financing new reactors through CWIP charges. CWIP charges are seen to again socialize the risks/costs of new reactor construction by forcing utility ratepayers to finance said construction by paying fees to cover construction costs *before* they are actually receiving the electricity they are paying for. This again shifts the costs and risks of default onto the public (in this case ratepayers) while the privately owned utilities retain control over the eventual profits. As with the analyses on the national level, this analysis not only facilitates the identification of existing injustices associated with CWIP charges and the public financing of nuclear facilities, but also enables the proposal of other principles of energy justice derived from struggles on the local level. Additionally, on the local level, antinuclear organization's claims regarding the dangers posed to local communities by nuclear facilities, including claims for enhanced safety and security at existing reactors, as well as claims for the permanent closure of unsafe facilities are examined. In this area claims drawing from the paradigms of recognition, representation, and distribution reveal the multidimensional injustices connected to everyday operation of nuclear facilities from the standpoint of local communities. The identification of these multidimensional injustices, once again, facilitates the identification of other principles of energy justice that derive from the social locations of affected communities.

METHODOLOGY AND DATA SOURCES

In order to answer these questions, a qualitative, discourse/textual analysis was performed on news coverage, press releases, action alerts, reports, legal interventions, flyers/leaflets, banners, blogs, letters to editors, op-eds., newsletters, art/music, petitions, "answer documents," and "platforms" associated with antinuclear SMOs on the local and national levels produced between March 2010 and August 2011. This eighteen-month timeframe encapsulates many important developments in nuclear energy policy in the United States, which resulted in large volumes of social movement activity and discourse. More specifically, this period of time extends from the cancelation of the Yucca Mountain project and the inception of the BRC (March 2010), through the release of the BRC's final report (July 2011), which continues to shape debates over federal radioactive waste policy. This eighteen-month timeframe also includes the movement's activities and claims during the first six months of the ongoing Fukushima nuclear disaster, which allows for an analysis of how this event impacted antinuclear social movement activity. It should be noted that this eighteen-month period is not representative of the "typical" amount or type of antinuclear social movement activity. There was far more activity in this period than in any other eighteen-month period that preceded it over the previous decade. However, for this reason, this period of time is especially illustrative of the breadth of antinuclear social movement activity, and examining these activities in this period allows for general conclusions about the movement on different analytic levels to be formulated.

Discourse is based on text (Johnson, 2002). More specifically, "discourse refers to the sum total of the 'manifestos, records of debates at meetings, actions of political demonstrators, newspaper articles, slogans, speeches, posters, satirical prints, statutes of associations, pamphlets, and so on' of a time, a place, and a people" (Sewell, 1980, pp. 8–9; quoted in Johnson, 2002, pp. 67). Thus, SMOs operating on different levels are the unit of analysis, with a focus on organizational discourse as it relates to broader, cultural discourses on energy policy, democracy, and justice. In social movement research, qualitative discourse analysis typically involves an intense focus on movement-related texts to identify patterns, linkages, and structures of ideas. This strategy is especially well suited to highlighting the deep structures of ideas and their relations within a movement, and to mapping the normative components of movement activity and formation (Johnson, 2002). Other research methods and data sources, such as ethnography and/or interviews, could be used to explore how *individuals* and *activists* within antinuclear SMOs construct and understand these issues. However, the interest of the present study is in the ways antinuclear SMOs behave and act as political/organizational actors in the course of political conflict with other political/organizational actors (i.e., the state, regulatory agencies, the nuclear industry, utilities, etc.) And because of the intensive focus on organizations' and the movement's textual materials, textual/discourse analysis is especially well suited for addressing "how questions"—how does the movement express its grievances; how does the movement construct and utilize frames and conceptualizations of justice; how does the movement utilize broader cultural symbols and representations. Therefore, this methodology proved to be an appropriate approach for answering the central research questions discussed earlier.

As is common in frame and discourse analysis, qualitative methods of data reduction, analysis, and presentation are employed. More specifically, a wide array of textual materials were organized by categories corresponding to the analytic levels of the movement. Next, these categorized textual materials were ordered by subcategories corresponding to the concepts of distribution, recognition, and representation. These coding categories were generated by the theoretical interests discussed earlier. The presentation of the findings below relies on texts that are presumed to be representative of the given category. "Intensive textual analysis must always balance its insights with the looming question of whether the text is representative enough to generalize about its patterns" (Johnson, 2002, pp. 71). The study seeks to strengthen the argument for the representativeness of the data, and the limited generalizability of my findings, through two strategies. First, the sample size is quite large for discourse analysis. It includes thousands of news reports, press releases, action alerts, reports, legal interventions, flyers/leaflets, blogs, letters to editors, op-eds., and so on, and the final dataset consisted of several hundred of these data. Second, the author has been an active participant in the antinuclear power movement for many years at both the local and national levels. This includes several years in the role of Radioactive Waste Policy Analyst for Missourians for Safe Energy (MSE), which is a local level, grassroots safe energy organization based in central Missouri (see below). The author has also given testimony at public hearings organized by the NRC on the issue of radioactive waste storage and attended the National Grassroots Summit on Radioactive Waste Policy in 2010. These personal experiences working in and with the antinuclear power movement enable the author to speak to the representativeness of specific examples in relation to broader movement discourse. Additionally, included in this large sample are several widely distributed and/or "milestone" documents, including movement "platforms" and "answers documents" relating to the Blue Ribbon Commission's "key questions"—both of which have been signed by hundreds of organizations (on both levels) and thousands of individuals.

The sampling procedure was both strategic and theoretical, reflecting the study's interests in both discourses of social justice in the antinuclear movement and the identification of principles of energy justice. On the national level, the analysis focuses on the five largest and most active antinuclear SMOs—NIRS, Friends of the Earth (FOE), Beyond Nuclear, Physicians for Social Responsibility (PSR), and the Union of Concerned Scientists (UCS). These national level organizations not only work closely with one another but also represent nearly the full spectrum of professional, dedicated, D.C.-based antinuclear organizations. For the local level, the analysis strategically relies on active SMOs that represent both different regions of the United States and also those who focused on different aspects of nuclear power. More specifically, local level SMOs working in the Southeast, Southwest, and Midwest were selected, as these are the regions of the country with the greatest concentration of nuclear industry activities going back to the Manhattan Project (Kuletz, 1998). Additionally, SMOs who primarily work on different aspects of the nuclear power issue were included in the analysis; such as increasing the safety of operating nuclear reactors, closing unsafe reactors, and fighting the construction of new reactors. These strategic sampling approaches helped ensure the findings on the local level are as representative as possible. Table 0.2 provides a basic overview of some of the antinuclear organizations considered for this study.

Through an analysis of the claims made by these and other SMOs on the national and local levels of the antinuclear power movement in the United States, the goal of the present study is to shed new light on the multidimensional and interrelated injustices associated with the production of nuclear power, in order to identify principles of energy justice that should guide future energy policy and decisions in a more socially and environmentally beneficial direction. The theoretical and methodological approaches taken here enable the exploration of the contours and deep interconnections between the ideational constructs that inform/underlie social movement activity, and how these can illuminate features of social, environmental, and energy justice. These new insights and types of knowledge will become increasingly salient as the realities of global warming become more difficult to deny, resulting in a forced restructuring of energy policy in the United States and around the globe. Practically, the findings of this research are relevant to the movement for a nuclear-free energy future itself. The analysis of the discourses of justice the antinuclear movement employs on the national and local levels can help

Table 0.2 Some of the Antinuclear Power Social Movement Organizations Analyzed in the Present Study

Antinuclear Organization Name	*Level of Operation*	*Organizational Focus and Campaigns*
Beyond Nuclear	National (United States)	All aspects of nuclear policy on the federal level, including nuclear development, radioactive waste management, and nuclear safety.
Friends of the Earth (FOE)	National (United States)	Work to reduce risks for people and the environment by supporting efforts to close existing nuclear reactors and fighting proposals for new reactors.
Nuclear Information and Resource Services (NIRS)	National (United States)	All aspects of nuclear policy on the federal level, including nuclear development, radioactive waste management, and nuclear safety.
Physicians for Social Responsibility (PSR)	National (United States)	Focus on educating the public about and eliminating the risks of ionizing-radiation that is associated with the production of nuclear power and radioactive waste.
Union of Concerned Scientists (UCS)	National (United States)	Focus on educating the public about and eliminating the risks of ionizing-radiation, as well as the engineering problems at existing nuclear sites.
Citizens Action Network (CAN)	Local (NY, VT, Mass., CN, NH)	Various grassroots, community organizing/empowering campaigns in New England to eliminate nuclear facilities in the Northeastern United States.
Indian Point Safe Energy Coalition	Local (NY)	New York-based organization focused on closing Indian Point NPP, which is the same make/model as those at Fukushima and is 25 miles from New York City.
Missourians for Safe Energy (MSE)	Local (MO)	Missouri-based safe energy organization formed to oppose the construction of nuclear reactors in the state, and opposes the ratepayer financing of new reactors.
Nuclear Watch South	Local (GA)	Georgia-based organization formed to oppose the construction of nuclear reactors in the state, and engages in nonviolent, direct-action across the Southeast.
San Lois Obispo Mothers for Peace	Local (CA)	California-based organization that emerged from the Abalone Alliance, and has continued to oppose the Diablo Canyon NPP.

Table 0.2 (Continued)

Antinuclear Organization Name	*Level of Operation*	*Organizational Focus and Campaigns*
Shut It Down Affinity Group	Local (VT)	Vermont-based organization concerned with closing the Vermont Yankee NPP.
Southern Alliance for Clean Energy	Local (FL, GA, NC, SC, TN)	Various grassroots, community organizing/empowering campaigns in the Southeastern United States to eliminate nuclear facilities.
Vermont Yankee Decommissioning Alliance	Local (VT)	Vermont-based organization concerned with closing the Vermont Yankee NPP.

generate new strategies for movement coordination between these levels. The findings presented below will similarly help identify which discourses and/or combination of discourses are most strategically useful, both in terms of activist mobilization and effect on public policy. Additionally, the findings will be useful for identifying which other broad-based social movements the antinuclear movement might form strategic coalitions with.

The present study's focus on the normative as well as the empirical basis of social movement activity might seem problematic and speculative to more empirically oriented scholars. However, as the earlier discussion indicates, in a pluralist knowledge culture the empirical and normative are inextricably interdependent. The interest in the differing conceptualizations of social justice that inform social movement activity necessitated the use of qualitative discourse analysis, which is well suited for addressing questions of how the movement expresses its grievances, how the movement mixes and combines ideologies and frames, how the movement utilizes broader cultural symbols, and so on. By virtue of its intensive textual focus, however, this methodological approach can only secondarily and/or implicitly take up central "why questions"—why does a particular movement succeed or fail, why does a movement take a certain trajectory, and so on. While these questions are undoubtedly important, they fall outside the focus of the current project.

Chapter 1

Federal Financing of New Nuclear Reactor Construction

In February 2012, President Obama announced his FY 2013 budget request, which for the first time since 2009 did not call for the expansion of the loan guarantee program for new nuclear reactor construction. In his FY 2011 and FY 2012 budget requests, which were made during the period of time under examination, President Obama called for a major expansion of the nuclear loan guarantee program. In fact, in both years President Obama specifically endorsed the expansion of nuclear power in the United States during his State of the Union Addresses, and then requested the loan program be expanded by $36 billion on top of the $18.5 billion allocated already. The original $18.5 billion was authorized by the Bush administration in 2007, and the anti-nuclear movement has since dedicated itself to stopping the expansion of the program, and ultimately canceling the program altogether. Thus, the Obama administration's cancelation of the $36 billion nuclear loan guarantee *expansion* in the FY 2013 budget request is viewed as a "huge grassroots environmental victory," albeit a defensive victory (Beyond Nuclear, 2012a). This victory was viewed by the antinuclear movement as being the result of almost five years of coordinated grassroots actions targeting the president, Energy Secretary Chu and his predecessors, and members of Congress who sit on related committees. Nuclear Information Resource Services (NIRS), Beyond Nuclear, and other national-level SMOs have organized the sending of tens of thousands of letters and thousands more phone calls to elected officials.

The issues of federal (taxpayer) financing of new reactor construction are perhaps the most direct ways to see the distributive, as well as some of the recognitional and representative injustices associated with the continued development of nuclear energy. Between March 2010 and August 2011 there were several attempts by the president and members of Congress to expand the nuclear loan guarantee program. All of these attempts were met with organized resistance from the antinuclear movement. More specifically, antinuclear organizations drew from the social justice paradigms of recognition, representation, and especially distribution to argue against the federal "loan guarantee" program by highlighting the multidimensional injustices associated with the program. It is necessary to identify and correct these multidimensional injustices in order for socially, politically, economically, and environmentally just energy systems to be financed and developed. Through an analysis of the claims made by antinuclear organizations regarding the multidimensional injustices involved with the federal "loan guarantee" program, it is possible to identify principles of energy justice that should guide future decisions about the development and financing of energy systems. The following analysis shows energy justice requires that the development of energy systems should minimize up-front, capital-intensive requirements to the greatest extent possible. Additionally, energy justice requires that in cases where public financing of energy systems is necessary, because private financing is unavailable for whatever reason, the eventual profits/benefits should be retained publicly. Beyond these primary analytic contributions, the following analysis also secondarily shows how the antinuclear power movement's multiple framings of the issues (that are occasioned by the multidimensional nature of the injustices associated with nuclear power) are strategically vital resources, which enable the formation of coalitions with other organizations not generally concerned with the production of nuclear power. Furthermore, the analysis reveals the antinuclear movement would benefit from expanding the implied subject position of *who* is entitled to different dimensions of social and environmental justice. This is especially the case with the associations found between the distributive injustices of the "loan guarantee" program and the subject position "taxpayers," on the one hand, and the recognitional and representative injustices associated with the "dirty and dangerous" aspects of nuclear power and subject positions of "communities" and/or "citizens," on the other.

ROUND ONE (APRIL AND MAY 2010): $9 BILLION AUTHORIZATION IN AN "EMERGENCY SUPPLEMENTAL FUNDING BILL"

During the crucial time period under consideration here, the antinuclear movement's claims regarding the maldistributive aspects of the federal "loan guarantee" program for new reactor construction began immediately. Within this time period, the first activities regarding the expansion of the loan guarantee program for new nuclear reactors took place in May of 2010. In February 2010, President Obama announced the first actual nuclear loan guarantee himself after instructing the DOE to award $8.3 billion in loan guarantees to Southern Company to help build two new reactors at plant Vogtle in Georgia. At this point in time, however, Southern Company and its subsidiary Georgia Power and Electric Co. had yet to accept the loan guarantee, and were at least two years away from receiving a license from the NRC. At this juncture, it was clear to antinuclear groups that the loan "guarantees" were not just guarantees but rather direct loans from the U.S. Treasury Department, with the actual money coming from the little-known Federal Financing Bank (NIRS, 2010d).

The $8.3 billion in loan guarantees awarded to Southern Company left $10.2 billion available from the $18.5 billion that were authorized in 2007 for the construction of new nuclear reactors. However, because the estimated cost of new reactors had continued to soar, the DOE realized the $10.2 billion would only cover one additional project. Initially, the rational for the loan guarantee program was to provide federal funds ($18.5 billion) for the construction of *five or six* new reactors, which was believed would stimulate private investment in new reactor construction. This tremendous growth in the expected costs of new reactors is why the Obama administration and the DOE were pushing to triple the loan guarantee program in the following year's budget to $54.5 billion. Apparently waiting for the following year's budget to be authorized was not acceptable to the DOE who seemed intent on maintaining the momentum generated by the loan guarantee award to Southern Company. So on April 28, while testifying before Congress, Energy Secretary Chu argued there was not enough "ceiling" in the current nuclear loan guarantee program and asked for $9 billion (out of the additional $36 billion being sought) to facilitate the awarding of two more loan guarantees in the coming weeks. The result was a $9 billion authorization that was attached to an emergency supplemental funding bill intended to provide funds for U.S. troops in Afghanistan and Iraq and to provide additional disaster relief money (related to the BP Gulf oil spill).

On May 26, 2010, the day before the House Appropriations Committee was set to consider the bill, NIRS sent out an Action Alert to their supporters. This Action Alert encouraged NIRS supporters to contact their representatives and ask them to do everything in their power to remove the nuclear loans, especially NIRS supporters whose representatives were on the House Appropriations Committee. NIRS was quick to point out the new loan guarantee authorization was being included in an "emergency" supplemental funding bill, even though there was clearly no "emergency" requiring new nuclear loans. For instance, the then most likely candidates to receive the next loan guarantees (Unistar's project at Calvert Cliffs, MD, and Nuclear Resource Group's project in Southwest, TX—both of which are now defunct) were at least two years away from a license from the NRC, and utilities cannot actually receive any loan monies without a construction and operating license (COL). As far as NIRS and other antinuclear organizations were concerned, there was clearly no emergency here, except for perhaps the DOE's desire to score points with the nuclear industry (NIRS, 2010d). More specifically, NIRS repeatedly referred to the loan guarantee program as a "bailout" of the nuclear power industry, playing off the public's continued outrage over the federal bailouts for the financial industry in 2008/2009. The NIRS Action Alert stated the goal of the campaign was to send two clear messages to Washington:

> Stopping this $9 billion deal would send two clear messages: 1) emergency funding bills should not be used for non-emergency pet projects, especially not dirty and dangerous new nuclear reactors. It is obscene to even attempt to put these loans on an emergency bill like this; 2) not only is the $9 billion unacceptable, so is the full $54.5 billion the administration wants next year! (NIRS, 2010d)

From this Action Alert, several dimensions of the maldistributive injustices associated with the "loan guarantee" program become apparent. Not only is it "obscene" to attach new nuclear loan authorizations to an "emergency" funding bill, but the entire rational for nuclear loans violates principles of distributive justice. In terms of the maldistribution of financial risks, the concern arises from the fact that capital-intensive development projects, such as the construction of new nuclear reactors, face a significant risk of default on loans they receive. In the event an investor-owned utility (such as Southern Company or its subsidiary Georgia Power and Electric) defaults on a federal loan guarantee awarded for the construction of a new reactor, taxpayers would be left footing the bill. This is especially problematic considering both the Congressional Budget Office in 2003 and the Government Accountability Office in 2008 estimated the average default risk on a federal loan guarantee for new reactor construction could be as high as 50 percent. This is why Wall Street refuses to invest in nuclear power; they understand the uncertainty surrounding nuclear power construction makes such investments too risky. For instance, in 2007 six top investment firms told the DOE they were unwilling to finance new reactors in light of the industry's horrible financial track record (New York Times, 2010). Therefore, in the absence of private investments, utilities seeking to build new nuclear reactors must find some way of socializing the risk while retaining private control of any future profits. On the national level, this is accomplished through the federal loan guarantee program for new nuclear construction.

Hours before the House Appropriations Committee was set to consider the emergency funding bill containing the nuclear loan authorization, the meeting was unexpectedly postponed until after the upcoming Memorial Day recesses. It is unclear why the meeting was postponed; however, it is clear that NIRS' mobilization was having an impact. According to a NIRS Action Alert sent on May 28, many of the members of the Appropriations Committee were unaware of the nuclear provisions that had been snuck into the bill. Additionally, many members of the committee may not have been aware that what appeared to be only a $90 million authorization—due to the byzantine Congressional budgeting procedures—actually translated into a $9 billion loan for the nuclear industry with the funds coming straight from the Treasury (NIRS, 2010e).

On June 16, 2010, following Congress' Memorial Day recess NIRS sent another Action Alert informing their members that the committee had yet to reschedule the meeting. NIRS's position was; the longer that the committee delayed, and the more controversial NIRS and their supporters could make the loan guarantees seem to the committee, the more likely the whole idea would be scrapped (NIRS, 2010f). NIRS stated over 5,000 people had sent letters of protest to their representatives, which was contributing to the delay in moving the bill forward. While in terms of absolute numbers 5,000 letters is not a huge volume, compared to the millions of letters sent in protest of the Keystone XL Pipeline, for example. However, for representatives on the Appropriations Committee to receive 5,000 letters of protest regarding an appropriations bill is significant, as these are usually bills that do not demand a lot of public attention. In the Action Alert, NIRS reiterated their position from earlier Action Alerts arguing "Not only are taxpayer loans for new reactors not an emergency—they should be stopped entirely!" (NIRS, 2010f).

From the very outset of the period of time under consideration for this study, the antinuclear power movement's arguments against the federal "loan guarantee" program were based on claims regarding distributive (in)justice. This program is seen to be yet another mechanism for socializing the risk while privatizing the eventual profits from large development projects, which is an unjust distribution of financial benefits and burdens. In other words, having taxpayers assume the financial risks of new reactor construction, while investor-owned utilities retain control of the profits the reactor might eventually generate, unjustly distributes the risks and rewards of such capital-intensive development projects. This begins to suggest at least two potential principles of energy justice, which are further supported by my analysis of subsequent rounds of conflict over the "loan guarantee" program (discussed below).

First, energy justice requires that the development of energy systems should minimize the lifetime costs of building, operating, and maintaining energy infrastructure to the greatest extent possible. Lifetime costs are established through life-cycle assessment (LCA) that considers the full range of costs associated with a product or facility, which come from raw material extraction through materials processing, construction, operation, and maintenance. In keeping with very basic market principles, energy systems should be pursued in accordance with the least-cost options that are available. This is also captured by levelized costs of electricity (LCOE) analyses, which measure the net cost of electricity generation from a generating plant over its lifetime. This, of course, involves an honest calculation of the full costs of different energy systems, including the *externalized costs* associated with traditional coal, oil, and gas production, which levels the playing field for renewable energy sources like wind, solar, and geothermal. So while fuel-efficient vehicles and energy-efficient buildings and homes have higher initial, up-front costs, LCA and LCOE assessments show they are in fact the lowest-cost options over the long term. With regard to energy systems, LCA and LCOE analyses show nuclear energy is by far the most expensive source of electricity, and renewables are increasingly competitive with traditional fossil fuels (U.S. EIA, 2021).

Additionally, these calculations should be done without consideration of the "sunk costs" associated with energy companies existing physical infrastructure, because the decisions regarding the development of these infrastructure were not made democratically nor were they made to advance social and environmental well-being, which is the entire purpose of establishing principles of energy justice. The history of the fossil fuel industry generally (Malm, 2016), and coal (Scott, 2010) and petroleum (Freudenburg and Gramling, 2012) specifically, shows the fossil fuel infrastructure was not established in a just or democratic manner. On the contrary, the fossil fuel infrastructure developed through the nineteenth and twentieth centuries was made possible by the systematic misrepresentation and misrecognition of public concerns and interests. This consequently resulted in the inequitable, distributive injustices evidenced today by crippling poverty in Appalachian and Gulf Coast states, and multi-billion dollar fossil fuel companies whose fortunes are made by exploitation of these people and places. Thus the exclusion of the "sunk costs" associated with energy companies existing *physical* infrastructure does not extend to the people and communities that formed around these infrastructure. The *social* infrastructure (consisting of the people and communities dependent upon the fossil fuel economy) must be provided with a meaningful voice and educational opportunities and retraining for good jobs in the renewable energy economy. These people and communities cannot just be abandoned with the historical and on-going damages inflicted upon them unrecognized. The ideas informing this principle of energy justice also inform the Green New Deal proposal and the broader climate justice movement.

Second, energy justice requires that in cases where public financing of energy systems is necessary, because private financing is unavailable for whatever reason, the eventual profits/benefits should be retained publicly. This could take the form of dividend payments to taxpayers/ratepayers or future credits toward their energy costs. This principle of energy justice is meant to prevent the neoliberal transfer of public (i.e., taxpayer) money into private (i.e., corporate shareholder) hands. Financial returns on investments need to be commenserate with the level of financial risk involved in making the investment. If private energy companies and/or investor-owned utilities shift the financial risk of new reactor construction onto taxpayers (or ratepayers as discussed in chapter 5), then the taxpayers (or ratepayers) should be entitled to the financial returns. This excludes programs and policies such as tax rebates for the purchase of electric vehicles and other subsidies for the construction and installation of renewable energy production as these sorts of incentives do not socialize the financial risks while privatizing eventual profits. Both of these potential principles of energy justice should be used to guide future policy to ensure the most cost-effective energy choices are made, and to ensure public interests are not sacrificed in order to advance private financial gains. Furthermore, the following analyses of subsequent rounds of conflict in this area provide further support for these principles.

ROUND TWO (JUNE AND JULY 2010): $36 BILLION IN THE FY 2011 ENERGY AND WATER APPROPRIATIONS BILL

The antinuclear movement's claims regarding the distributive injustices of the "loan guarantee" program were reiterated in the following months, as were additional concerns about representative and recognitional injustices. The stakes of the conflict over the "loan guarantees" were raised the following week as reflected in the headline of another NIRS Action Alert sent on June 22: "Correction!!!!! House Appropriations Committee to meet Thursday afternoon—June 24—to consider $36 billion in taxpayer loans for new nuclear reactors. Last chance to act and mobilize!" NIRS had just learned the Obama DOE was requesting the entire $36 billion in additional loan authorizations, instead of waiting to include the request in the following year's budget. NIRS urged its members to take action immediately and release an advertisement about their campaign against nuclear loans that ran on Alternet and other progressive sites. On June 24, the House Appropriations Committee's Subcommittee on Energy and Water again abruptly canceled their meeting to consider the following year's energy budget—possibly including the additional $36 billion the DOE was requesting for nuclear loans. This was the second such cancelation within a month. While it was unclear at the time, the cancelation may well have been related to the over

11,000 letters sent by this point to Congressional representatives opposing the expansion of the loan guarantee program (NIRS, 2010h).

As revealed in the following week, the meeting of the Energy and Water Subcommittee was in fact canceled because of the nuclear loan issue. Apparently, even though the Obama DOE requested the additional $36 billion, the bill that was brought before the subcommittee did not include any nuclear loan money. This was largely due to committee chairman David Obey and Rep. Donna Edwards who played key roles in keeping that $36 billion out of the bill, and NIRS members in their districts were encouraged to write and thank them (NIRS, 2010i). Ultimately the subcommittee's meeting was canceled because other members wanted the nuclear loan money and objected when they discovered it was not included in the draft bill. The most vocal about their objections were Democratic representatives Chet Edwards (Texas), Chaka Fattah (Penn.), and Marion Berry (Ark.). The Action Alert called for people in their districts to "flood their offices with phone calls! Tell them to stop wasting taxpayer money on the wealthy nuclear power industry" (NIRS, 2010i).

Representative Edward's steadfast opposition to the bill (because it lacked nuclear subsidies) is interesting considering Edward's was then in a tightly contested race in Texas's seventeenth district, where he had turned the nuclear loan program's expansion into a central campaign issue. More specifically, Edwards supported the expansion of Nuclear Resource Group's South Texas project in his district, and had begun criticizing his republican opponent for suggesting that federal money should not be used to construct nuclear reactors. According to Ben Schreiber of Friends of the Earth Action,

> It is interesting that a $25 billion preemptive bailout for the nuclear industry appeared in the Energy and Water Appropriations bill shortly after it became a campaign issue for Congressman Chet Edwards. It's enough to ask whether the House leadership is risking $25 billion of taxpayer money so one House seat stays in the Democratic column. (Friends of the Earth, 2010)

This coincidence appeared to Friends of the Earth as signaling House Democrats were using the DOE loan guarantee program like a campaign credit card. Moreover, this was not an isolated incident of *misrepresentation* and undemocratic practices associated with the recent attempts to expand the loan guarantee program. The $9 billion for new reactor construction being considered with the emergency supplemental funding bill included funds for nuclear projects in then-Majority Leader Steny Hoyer and Whip James Clyburn's districts (Friends of the Earth, 2010). For the antinuclear movement these examples clearly show how the *maldistributive* injustices of the nuclear loan guarantee program are compounded, and in fact facilitated, by the *misrepresentation* and undemocratic decision-making procedures that accompany nuclear development.

The antinuclear movement's arguments that taxpayer-backed loan guarantees for new reactors is founded upon and advances distributive injustices took on a new dimension in early July 2010, along with the movement itself gaining new (somewhat surprising) allies. On July 1, NIRS sent out a press release with the headline:

> Foreign Companies, Workers Are Big Early Winners Under Federal Loan Guarantees for Nuclear Reactors: *Taxpayer-Backed Nuclear Loan Guarantees Make July 4th "Energy Non-Independence Day"; 2 of 3 Leading Loan Guarantees Candidates Have Major Foreign Ownership, Large Reactor Component Construction Jobs Are Overseas.* (NIRS, 2010j)

This press release related to a new report prepared by NIRS titled "Nuclear Power: Implications of Loan Guarantees for Reactors with Foreign Control and Foreign Jobs." In this report, NIRS documents how many of the then eighteen currently pending reactor projects in the United States involved reactors that were designed by either French or Japanese companies. Additionally, the report showed the immediate employment benefits of the loan guarantees will flow to non-U.S. workers, since virtually all major reactor components are made outside of the United States by foreign companies (NIRS, 2010j). More specifically, the Calvert Cliffs project in Maryland was slated to use a reactor supplied by France's AREVA, and the South Texas project would use a reactor supplied by Japan's Toshiba Corporation. Thus the loan guarantee program was an especially unjust distribution of economic benefits and risks, with the benefits flowing to foreign workers and foreign, state-supported nuclear companies and the risks being borne by American taxpayers.

Issues of Americentrism and xenophobia aside, it is interesting to note how the perceived problems with the U.S. loan guarantee program are situated with reference to the globalized nuclear supply chain. Michael Mariotte, then executive director of NIRS, summarized the maldistributive aspects of the loan guarantee program with reference to the globalized nuclear supply chain:

> The bailout of the nuclear industry with taxpayer-backed financing of loan guarantees is in no way a triumph for U.S. energy independence. Instead, it is a huge publicly backed corporate welfare arrangement for foreign-owned companies and non-U.S. workers. The non-U.S. companies that stand to be the biggest beneficiaries of taxpayer-backed loan guarantees are both massive in size and profitable. If American taxpayers were upset about bailing out U.S. banks and car companies, they should be furious about being put at risk in order to fatten the bottom line of overseas nuclear companies. (NIRS, 2010j)

Interestingly, NIRS was joined in releasing the new report by a senior/intergenerational group—the Alliance for Generational Equity (AGE)—and a taxpayer's advocacy group—the Association of Concerned Taxpayers (ACT)—to oppose taxpayer-backed loan guarantees. The press release explicitly stated AGE and ACT *do not* oppose privately financed nuclear power; however, they *do* strongly object to taxpayer-backed loan guarantees. Therefore, by limiting their framing of the problem to the maldistributive aspects, NIRS was able to form a coalition with these relatively conservative organizations. In the press release David Herman, vice president of AGE, said:

> This is not about whether you support or you oppose nuclear power. . . . Whether it's for banks, car companies, investment firms or nuclear reactors, a bailout is a bailout is a bailout. We need a little less socialism from Washington and a lot more faith in the wisdom of the marketplace. . . . If these companies controlled by French and Japanese interests want to finance new reactors and find investors who want to support that, they have our blessing. But we are not prepared as U.S. taxpayers to allow seniors and other Americans to be ripped off by yet another multi-billion-dollar bailout for another industry. (NIRS, 2010j)

Gordon Jones, president of ACT, expressed similar sentiments in the press release:

> Taxpayer guarantees of private economic actions represent the same kind of implicit subsidy that caused the mortgage meltdown of 2008. Such subsidies encourage behavior that is riskier than unsubsidized individuals and entities would undertake without the guarantees. . . . ACT favors a free market in energy, without government intervention for or against any particular form of power. . . . Decisions about energy production should be made by market forces, not by politicians in Washington or by bureaucrats at the Department of Energy. (NIRS, 2010j)

It is clear from these statements that AGE and ACT are concerned with only the maldistributive aspects of the loan guarantee program and not with the misrecognition or misrepresentation associated with the production of nuclear power and radioactive waste. Therefore, although both groups oppose the federal loan guarantee program for new nuclear reactors, they are not antinuclear social movement organizations. Rather, for both groups distributive justice results from (and is achieved through) the operations of a "free market." Using the language of "bailouts" and "subsidies," both AGE and ACT claim it is unjust to risk American taxpayer dollars on projects that will benefit foreign workers and foreign companies.

As an organization AGE is primarily concerned with intergenerational equity in terms of the national deficit and debt and is motivated by concerns over future generations having to pay off deficits incurred today. AGE's framing of the issue as "another bailout for another industry" creatively links to the ubiquitous populist rhetoric/anger of the time, best expressed by the Tea Party movement. The basic argument being made is: here's another example of everyday people being made to financially support a large, profitable industry, which is unfair to present and future generations. The situation is considered especially problematic because the beneficiaries in this case are *foreign*-controlled companies and *foreign* workers. However, AGE's concerns for intergenerational equity appear to be limited to only concerns over deficit spending, as their neutral support for nuclear power would suggest they see no problem with leaving future generations tens of thousands of tons of high-level radioactive waste to manage.

Similarly, ACT organizationally promotes conservative tax and financial policies and advocates for a "free market" unencumbered by government regulation/interference. Their framing of the issue as a "subsidy" for the industry (one that distorts market signals and thus disrupts the operations of the market) again plays to the populist rhetoric of the time. Additionally, ACT's statement partially incorporates insights generated by ecological economists about "bad subsidies" encouraging risky/destructive behavior that otherwise would be discouraged by a "free market" (Foster et al., 2011; Speth, 2008). However, what is most interesting about ACT's position is the incorporation of specific ideas of just participation/representation that should be involved in the related decision-making processes. Opponents of the loan guarantee program primarily frame the problem as one of maldistribution, with alternate framings of the issue (i.e., misrecognition and misrepresentation) being utilized in an auxiliary manner. Here ACT is very explicate about who should make the decisions about energy production: "Decisions about energy production should be made by market forces, not by politicians in Washington or by bureaucrats at the Department of Energy." On the surface this might appear to some as being undemocratic, especially for many antinuclear organizations and activists that are generally very concerned with advancing democratic and other participatory forms of decision-making. But from the neoliberal worldview this group inhabits, freedom of market forces is absolutely synonymous with individual freedom, liberty, and democracy.

Despite the subtle difference between the positions of AGE and ACT, both members of the nascent coalition are potentially important allies for the antinuclear movement. These organizations are composed of (and have influence over) people who otherwise may not be concerned with the politics of nuclear power. AGE and ACT are both organizations that generally share little in common with antinuclear organizations such as NIRS, Beyond Nuclear, and others. However, because of the multiple and complementary framings of the issue used by the antinuclear movement, "frame alignment" between these groups was possible with regard to problems of maldistribution (Snow and Benford, 1986, 1992). This shows the strategic advantage of utilizing multiple and complementary discourses of social and environmental justice in the course of environmental justice energy activism. In different contexts, different discourses of social and environmental justice can be deployed and reconstituted to advance the movement's interests. In different ways, both AGE and ACT expand the discourse of maldistribution surrounding the loan guarantee program, and in so doing expand the potential audience for the antinuclear movement's message.

Despite the report released by NIRS, AGE, and ACT on July 1, 2010, the House of representatives passed the emergency supplemental funding bill, including the provision for $9 billion for new loans for new reactor construction. This was despite the more than 13,000 letters of protest NIRS supporters had sent to their representatives and thousands more phone calls (NIRS, 2010k). According to a NIRS Action Alert sent out on July 2, 2010, there was no separate vote on the nuclear loans, as they were included in a broader domestic spending package which was approved by a 239-182 votes. Most Democrats voted for the package, most Republicans voted against it. At this point the movement's attention turned to the Senate, because it earlier passed a different version of the bill (without the nuclear loans), and the Senate then had to decide whether to accept the House version. On July 22, the Senate rejected the House-passed emergency supplemental funding bill by a 51-46 votes. Again, most Senate Republicans voted against the bill, not because they oppose nuclear power, but because they are opposed to domestic spending more generally (NIRS, 2010l). According to a NIRS Action Alert sent on July 23, the House would surely have to accept the Senate's version of the bill and "these loans [were] now history" (NIRS, 2010l).

What is most significant about this round of conflict over the "loan guarantee" program is it clearly shows how the multidimensional injustices of the program enable the antinuclear power movement to frame their grievances in different terms, alternately focusing on issues of distribution, recognition, and representation. These multiple frames of their grievances can be a strategically vital resource that enabled the movement to form a coalition with other groups who are generally not concerned with issues of nuclear power. More specifically, by focusing on the maldistributive aspects of the program, the antinuclear movement was able to achieve "frame alignment" (Benford and Snow, 1986) with fiscally conservative groups who only oppose the public financing of development projects. Despite the very different concerns and perspectives of groups such as ACT and AGE from those of antinuclear organizations, the above analysis further supports the principles of energy justice proposed above. Again, through an analysis of the claims regarding the multidimensional injustices associated with the "loan guarantee" program, it appears that energy justice requires advancing the most cost-effect options available for energy development (assessed over the entire life-cycle of the facility) and that when public financing is required for such development projects, the public should retain control over the eventual benefits and/or profits.

ROUND THREE (JUNE–JULY 2010): $9 BILLION IN *ANY* BILL THAT COULD CLEAR THE SENATE

Whatever satisfaction the antinuclear movement and other opponents of the loan guarantee program were feeling about winning another round in the fight was short lived. A week later on July 30, 2010, NIRS sent another Action Alert reporting Senator Hutchison (R-Texas) was trying to attach the $9 billion authorization to any bill that might get through the Senate. For instance, Senator Hutchinson offered it to a bill to assist small businesses, despite the fact that no nuclear company (manufacturer, utility, etc.) could even remotely qualify as a small business (NIRS Action Alert, 7/30/10). Majority Leader Reid, however, refused to let this amendment take up valuable Senate floor time. At this point, the DOE was under perceivable pressure to grant loan guarantees for Unistar's project in Maryland and/or NRG's project in South Texas. Representatives from both firms had already publicly announced their needs for a loan guarantee in the very near term, or both projects faced the possibility of collapse (NIRS, 2010l).

Reflecting the general unproductivity and gridlock that characterized the U.S. Senate in this period of time, nothing more about the expansion of the loan guarantee program for new reactors was seriously considered until following the midterm elections in November 2010. Of course, the 2010 elections evidenced the rise in power and influence of the Tea Party movement. This is especially true considering the Tea Party's success in several high-profile House races, which in part resulted in Democrats losing control of the House. Antinuclear organizations and activists were initially unsure what this would mean for the loan guarantee program. On the one hand, Republicans have long supported nuclear power. High-profile Republican Senators such as Lamar Alexander (TN), Orin Hatch (UT), Mitch McConnell, Richard Shelby, and others have long advocated for a dramatic expansion of nuclear power in the United States and around the world. Moreover, it was a Republican-controlled Congress that passed the 2005 Energy Act, establishing the loan guarantee program and securing the program's initial authorizations. However, the Tea Party movement represents a (slightly) different brand of conservatism. As discussed above, by the end of 2010 a growing number of groups who saw themselves as fiscal conservatives were arguing along with antinuclear groups that loan guarantees unfairly shifted the risks of default to taxpayers, and government money could be better spent elsewhere. Conservative think tanks such as the Heritage Foundation and other fiscal-conservative groups like the National Taxpayers Union, Taxpayers for Common Sense, the Competitive Enterprise Institute, and the Non-proliferation Policy Education Center all signed a letter to Congress on December 8 that called for halting any increase in the amount allocated to the loan guarantees for nuclear plants (International Business Times, 2010). In short, several of the incoming representatives ran on a platform of fiscal restraint and limiting the kind of government interventions into the market that the loan guarantee program embodied.

ROUND FOUR (DECEMBER 2010): $9 BILLION IN THE "CONTINUING RESOLUTION"

Before these issues concerning the new Congress could become relevant, however, the DOE initiated another round of controversy surrounding the loan guarantee program during the "lame duck" session. This time the Obama DOE was asking Congress for the $9 billion for additional taxpayers loans as a part of the upcoming "Continuing Resolution" that was needed to keep the government funded (NIRS, 2010n). This was at least the fourth attempt in 2010 to increase the amount of money available for new reactor construction. NIRS reiterated their position arguing the "loan guarantees" were not "guarantees" but actual *loans*, with the money coming from the Treasury. Not only that, but the money was being given to large, profitable, *foreign* companies such as Toshiba and Electricite de France (EDF), which is the largest electric utility in the world and heavily subsidized by the French State. A NIRS Action Alert rhetorically asked, "When will Congress learn that taxpayers don't want to subsidize giant corporations, especially when they want to build dirty and dangerous new nuclear reactors in our communities!" (NIRS, 2010n).

Here again we see how frame convergence (Snow and Benford, 1986) is accomplished by situating claims about the dirty and dangerous aspects of nuclear power with reference to concerns about the maldistributive/expensive aspects. This shows how environmental justice energy activism not only involves multiple discourse/conceptualizations of social and environmental justice but also how these multiple discourses are strategically vital resources. Also note when the dirty and dangerous auxiliary frames (Snow and Benford, 1992) are utilized again the subject position of who is entitled to justice shifts from "taxpayers" to "our communities." On some level the association of the subject position "taxpayers" with the expensive/maldistributive frame and "our communities" with the dirty and dangerous frames makes sense given the nature of the two (interrelated) issues. However, it is also the case that "our communities" also suffer from the maldistributive aspects of the loan guarantee program. Many of "our communities" are struggling to fund basic social services such as education, infrastructure maintenance, and health services, among many others. Is it not unjust to "our communities" to lavish billions of dollars on profitable, foreign companies while these needs are left unmet? Additionally, "taxpayers" will surely be (and have been) victimized by the dirty and dangerous aspects of nuclear power. "Taxpayers" are exposed to the accidental and routine release of radioactivity from nuclear reactors, and will be put at risk by the transportation of high-level radioactive waste. In short, "taxpayers" will be subject to having their concerns misrecognized and their interests misrepresented as well. Expanding the discursively constructed subject position of *who* is entitled to social and environmental justice (with reference to all aspects of nuclear power) would advance the antinuclear movement's cause by broadening their potential audience.

The NIRS Action Alert from December 7 informed their supporters,

> With your help, we beat back the first three attempts to increase funding for this program. This one will be the hardest to beat yet, since the Administration's request is wrapped up in a larger package. So we need the loudest outcry possible. (NIRS, 2010n)

Despite the mobilization efforts of NIRS and other organizations, two days later on December 9, the House passed a Continuing Resolution that included $7 billion of new taxpayer money for nuclear reactors. As before, the antinuclear movement's best bet for stopping the expansion of the program was in the Senate, where in fact the program's expansion was stopped two weeks later. On December 23, NIRS releases an Alert with the headline: "Happy Holidays—Another Huge Victory! Government Funding Bill Enacted Without Nuclear Loans!" The Alert went on to say:

> THANK YOU! You've done it again! You sent more than 15,000 letters to Congress in December and made many, many phone calls to stop $8 billion in taxpayer loans for new nuclear reactor construction. And the final government funding bill, signed by President Obama, contains not one dime for new nukes! . . . That makes at least seven major efforts over the past two years by nuclear

> industry backers to increase taxpayer loans for new reactors—and every one of those efforts has been blocked! *Grassroots people power works!* (NIRS, 2010o)

ROUND FIVE (FEBRUARY–JUNE 2011): $36 BILLION IN FY 2012 BUDGET REQUEST

In the final round of the conflict over the "loan guarantee" program considered in the present study, the antinuclear movement not only again reemphasized their claims regarding the maldistribution of the program but also in response to the Fukushima disaster on March 11, 2011, began to explicitly connect these arguments to others drawing from the social justice paradigms of recognition and representation. These expanded arguments, and the discourses of recognitional and representative justice they draw from, further support the principles of energy justice proposed above, while suggesting at least one additional principle relating to the public (as opposed to private) financing of social and environmentally injurious energy systems.

Following the holiday recess, the issue reemerged in February 2011 with Obama's FY 2012 Budget request, which identical to the previous year's request included an additional $36 billion for new reactor "loan guarantees." A NIRS Action Alert sent on February 15, 2011, had the following headline: "Obama FY2012 Budget: Cut Heating Assistance for Poor, Cut Clean-Up of Great Lakes . . . But Shower Billions of Dollars in Loans to Wealthy Nuclear Utilities and Foreign Reactor Manufacturers!" (NIRS, 2011a). In addition to connecting the maldistributive aspects of the loan guarantee program to other pressing social and environmental problems, this Alert encouraged NIRS supporters to remind their representatives that in budget terms, $36 billion in loan "guarantees" shows up as only $360 million in actual spending. More specifically, the Alert reported:

> In the real world, those "guarantees" are actually taxpayer loans. The money leaves the federal treasury and only comes back if the project succeeds. During the first go-round of nuclear construction, fewer than half of the reactors proposed were actually ever built and generated income. . . . There is no reason to believe things will work out any better for the industry this time around—especially if we taxpayers are liable for the bill. (NIRS, 2011a)

On March 3, 2011, just over a week before the Great East Japan Earthquake and the onset of the Fukushima Diachii nuclear disaster, the battle over the loan guarantee program took a new turn in favor of the antinuclear movement. A major poll was released that day by the *Wall Street Journal/NBC News* that showed cutting subsidies for new reactor construction was the single most acceptable budget cut for the American public. More specifically, the poll showed 57 percent of Americans found cutting nuke subsidies was either completely or mostly acceptable, while only 20 percent think these cuts would be completely unacceptable. No other program areas mentioned in the survey got more than 52 percent, and most programs were under 50 percent (NIRS, 2011c). The antinuclear movement quickly seized the opportunity provided by this new poll data, with NIRS announcing:

> In this era of Congressional budget-slashing, with programs that actually benefit real people on the chopping block, we think it's time to step up our demands: Tell Congress not only must it oppose new taxpayer subsidies for new reactor construction, but Congress must cut the existing program to zero. (NIRS, 2011c)

For antinuclear organizations such as NIRS, the time seemed ripe for them to go on the offensive and demand the entire program be canceled and funds redirected. NIRS recognized one opinion poll was not going to convince Congress and implored their supporters to reach out to their representatives in order to make their voices as loud as possible (NIRS, 2011d).

Then came the events of "3/11," as it is known in Japan. On March 11, 2011, a 9.0 magnitude earthquake struck off the Northeast coast of Japan killing thousands of people and creating a more than forty-foot tsunami that devastated the Fukushima Diachii nuclear power facility, resulting in the ongoing meltdown of multiple reactors and the worst nuclear disaster since Chernobyl in 1986. In the resulting hectic days and weeks that followed, the world watched as reactor buildings exploded and industry and government officials scrambled to understand what was going on and inform/reassure a concerned public. Many antinuclear organizations were absolutely overwhelmed by the task of trying to accurately assess the situation in Japan. In the initial days and weeks that followed 3/11, seemingly nothing about what was happening at Fukushima was totally clear. This was due to both a dire *lack* of official, verifiable information coming from both Japanese or TEPCO officials and the *overabundance* of speculative reports exaggerating the consequences of the accident and/or baseless assurances that there was no real danger to the public or environment.

The Fukushima disaster affected most (but not all) discussions about nuclear power and radioactive waste in the United States. For the antinuclear movement, Fukushima changed everything—from their multiple and complementary framings of the issues to an emboldened and increasingly offensive strategy that no longer was content with defensive victories. As mentioned above, antinuclear organizations were overwhelmed during the first weeks following 3/11 trying to assess the situation and inform their members and the media. It was a tremendous accomplishment that nonprofit, grassroots organizations (like NIRS, Beyond Nuclear, and others) operating on shoestring budgets were able to so successfully wade through the noise and educate millions of people about the Fukushima disaster directly and through national and international media.

By April 13, 2011, NIRS had the opportunity to catch its breath and release an Action Alert asking their supporters to tell Congress, "The lesson of this catastrophe is clear: we must end the use of nuclear power. And that must start with the prevention of any new nuclear reactors" (NIRS, 2011e). The Alert goes on to argue,

> It is outrageous that the Obama Administration continues to say nuclear power will be part of its "clean energy" strategy and continues to seek $36 Billion *MORE* in taxpayer loans for new reactor construction. This funding must be stopped, and existing taxpayer subsidies for the nuclear industry withdrawn. How anyone can view the images coming from Japan and continue to claim nuclear power is somehow "clean" is beyond our ability to comprehend. *If ever there was a time for Congress to hear our voice, it is now.* (NIRS, 2011e)

In the first month following the onset of the Fukushima disaster more than 40,000 people had sent letters of protest to their representatives, calling for an end to the use of nuclear power in the United States (NIRS, 2011e).

In mid-May 2011, NIRS began circulating an "organizational sign-on letter" to end the nuclear loan program, which was ultimately signed by over 180 organizations and small businesses. More specifically, 183 organizations signed the letter, representing millions of people from Maine to Hawaii. These groups consisted of national-level organizations, such as Greenpeace, Friends of the Earth, and Clean Water Action, as well as local and regional environmental and peace groups, family farms, food co-ops, and other small businesses (NIRS, 2011f). On May 25, the letter was hand-delivered to members of the House and Senate Appropriations Committees on Energy, who were still considering the expansion of the nuclear loan program. In fact, on Day 5 of the Fukushima disaster (March 15, 2011) President Obama had again asked Congress for the additional $36 billion in new nuclear loan authorizations. For many organizations in the antinuclear movement this clearly signaled the administration's refusal to recognize both the seriousness of the Fukushima disaster and the now "obvious" dangers associated with nuclear power.

Consistent with earlier efforts, the letter asked members of the House and Senate Appropriations Committees to reject the new request for the additional $36 billion, as the expansions of the program represented an unjust distribution of scarce financial resources. Additionally, following Fukushima the antinuclear movement continued to go on the offensive by also asking that the unspent $10.2 billion remaining in the program be rescinded, as well as the $8.3 billion that had been conditionally offered to Southern Company for the plant Vogtle expansion. In short, the letter argued the Title 17 program for nuclear power should be ended immediately. According to the letter, the Fukushima disaster severely eroded the entire public policy rational for the program. When enacted in 2005 and funded in 2007, Congress believed "modest funding" ($18.5 billion) to support the construction of six to eight new nuclear reactors would encourage private investment—which even back then was extremely skeptical of investing in new nuclear projects. The letter goes on to argue:

> Since then, estimated construction costs have skyrocketed so that it would take more than $50 Billion to support construction of those same 6-8 new reactors. And Fukushima has provided a stark reminder to the investment community that a multi-billion dollar investments can turn into much larger liabilities overnight. [On March 10th, 2011] Tokyo Electric Power was one of the three largest electric utilities in the world. Today it is a shambles, facing tens—perhaps hundreds—of billions of dollars in liabilities. (NIRS, 2011f)

The letter then pointed out the only new reactors being constructed anywhere in the world were being built entirely with government funds, and the private investment community will not put its money into new nuclear power plants, no matter how much support is given through taxpayer dollars. Therefore, the letter concludes, "Title 17 cannot and will not lead to greater private investment in nuclear power and thus cannot and will not achieve its goals" and this alone should be reason enough to end the program *before* taxpayer money is put at risk (NIRS, 2011f). The letter then points to the fact that the American people agree, citing poll data from the Civil Society Institute that shows 73 percent of the American people oppose federal loan guarantees for new nuclear reactors, and the *Wall Street Journal/NBC News* poll that showed 64 percent of Americans oppose new reactor construction entirely. Finally, the letter points to the post-Fukushima decisions of Germany and Japan—the world's fourth- and third-largest economies respectively. These countries, it was argued, were exploring new energy policies that center on clean renewable energy and increased energy efficiency, and were moving away from nuclear power.

> These countries understand that clean energy is indeed the future and that clean energy plants do not explode and release toxic radiation across the world. The United States should join Japan and Germany and become a leader in clean energy rather than falling further behind in the race to implement the energy sources of the 21st century. . . . The Title 17 loan guarantee program for nuclear

> power unnecessarily risks billions of taxpayer dollars on a program that cannot meet its own goals and that simply diverts our limited resources away from the goal of safe, clean and affordable energy for our nation. (NIRS, 2011f)

This widely supported letter encapsulates the many dimensions of the antinuclear movement's efforts to fight the "loan guarantee" program during the period of time under examination here. Perhaps the most direct way to see the distributive, as well as some of the recognitional and representative, injustices associated with the continued development of nuclear energy is through the issues of federal (taxpayer) financing of new reactor construction. Additionally, through the analysis of the claims made by antinuclear organizations regarding the multidimensional injustices involved with the federal "loan guarantee" program, it is possible to propose principles of energy justice that should guide future decisions about the development and financing of energy systems. More specifically, the above analysis indicates energy justice requires that the development of energy systems should minimize lifetime costs (established through LCAs) of building, operating, and maintaining energy infrastructure to the greatest extent possible. In keeping with very basic market principles, energy systems should be pursued in accordance with the least-cost options that are available, as assessed through LCA and LCOE assessments. This, of course, requires an honest calculation of the full costs of different energy systems, including the *externalized costs* associated with traditional coal, oil, and gas production, which levels the playing field for renewable energy sources like wind, solar and geothermal. Additionally, these calculations should be done without consideration of the "sunk costs" associated with energy companies existing *physical* infrastructure, which was built and operates undemocratically and does not advance social and environmental well-being, which is the entire purpose of establishing principles of "energy justice." This does NOT extend to the *social* infrastructure that has developed around fossil fuel and nuclear facilities. Workers and communities that are dependent upon nuclear and fossil fuel facilities must be fully represented in decision-making, have their needs recognized, and receive a just distribution of resources and opportunities. Secondly, energy justice requires that in cases where public financing of energy systems is necessary, because private financing is unavailable for whatever reason, the eventual profits/benefits should be retained publicly. Again, it is unjust for taxpayers (or ratepayers) to assume the tremendous financial risks of new reactor construction, while private, investor-owned utilities, and/or energy companies retain any realized profits. Public benefits could take the form of dividend payments to taxpayers/ratepayers or future credits toward their energy costs. Both of these potential principles of energy justice should be used to guide future policy to ensure the most cost-effective energy choices are made, and to ensure public interests are not sacrificed in order to advance private financial gains.

Throughout the various rounds of the conflict between March 2010 and August 2011, the antinuclear movement's arguments against the loan guarantee program evolved and developed in a number of interesting ways. Different maldistributive aspects of the program were identified and critiqued by antinuclear organizations and other concerned groups. By linking their arguments against the maldistributive aspects of the program to critiques of government "subsidies" and "bailouts," these critiques routinely drew from the populist rhetoric/anger that was so common during this period of time. Additionally, the preceding analysis has begun to highlight how in the course of environmental justice energy activism, the antinuclear movement not only draws upon multiple and complementary discourse of social and environmental justice, but also how these varied discourses become strategically vital resources. On the one hand, the antinuclear movement utilizes and reconstructs different discourse of social and environmental justice because the injustices of nuclear power are themselves multidimensional, involving unjust practices of maldistribution, misrecognition, and misrepresentation. However, on the other hand, the movement's use of these diverse discourses of justice also enables the movement to be flexible and adaptable, as we see with the coalition between NIRS, ACT, and AGE.

The antinuclear movement's efforts to oppose the loan guarantee program for new nuclear reactor construction rely primarily on the discourse of distribution, with the discourses of recognition and representation being used in an auxiliary, complementary manner. As will be explored in the next chapter, with regard to radioactive waste disposal (the other central issue on the national level for the antinuclear movement in the United States) the discursive strategy is quite different. The problems of radioactive waste disposal are primarily framed as problems of misrecognition and misrepresentation, and issues of distributive justice are here used to support this dominant framing/discursive strategy.

Chapter 2

High-Level Radioactive Waste Management

From the early days of the Manhattan Project and the 1954 Atomic Energy Act (which relaxed restrictions on nuclear technologies and began the commercial production of nuclear power), radioactive waste has been seen as *the central problem* of nuclear power production by both critics and proponents of the technologies. Perhaps the most direct way to see the environmental injustice problems created by the production of nuclear power is to examine the issues of radioactive waste storage and disposal. Energy systems take on significant social dimensions as these systems are developed, contested, and implemented (Nye, 2001; Sze, 2005, 2007). Therefore, the decision about the production, distribution, and consumption of energy and electricity involves more than just technical/scientific considerations. This is clearly revealed by analyzing the "environmental justice energy activism" of groups working to highlight these social dimensions of energy systems, such as antinuclear organizations. The "unsurpassable social and political obstacles" that necessitated the canceling of the Yucca Mountain Project (YMP) (discussed here) relate back to deep, structural, and normative representational, recognitional, and distributive injustices that have plagued the radioactive waste management program in the United States since the very beginning. Therefore, in order for the nuclear waste crisis to be solved in a technically, socially, and politically/economically acceptable manner, it is necessary to identify and correct these multidimensional injustices. Following the methodology established by Fraser (2000, 2003, 2009), we can begin to understand the full structural and normative extent of both just and unjust social arrangements by examining claims made in existing struggles against existing injustices as the starting point for the development of theories of justice (including energy justice).

Initially the Atomic Energy Commission (AEC) and its industrial contractors (e.g., G.E., Westinghouse, Union Carbide, and others) viewed the radioactive waste problem as a "solvable problem," and were primarily concerned with quieting what they saw as "unreasonable public hysteria" (Walker, 2009). Following the inception of the "Atoms for Peace" program in 1954, the burgeoning nuclear power industry spent almost as much time, resources, and effort on public relations as it did on technical research and development (Walker, 2009). By the mid-1960s the overpowering optimism within the nuclear industry and AEC that a "solution to radioactive waste storage was well on its way" gave way to the realization that radioactive waste had become a "huge and ever increasing problem" (Walker, 2009). By the late 1960s to early 1970s, a global consensus (among nuclear states) was reached that geological disposal was the best and only long-term solution for the disposal of high-level radioactive waste. However, in the United States the conflict of interests created by the dual mandate of the AEC (to both promote *and* regulate commercial nuclear power) began to give rise to a series of controversies involving the commission and their handling of radioactive wastes.

Public protest over the handling, storage, and disposal of radioactive waste began in the late 1950s with regard to ocean dumping—a practice that was quickly abandoned due in large part to public outcry (Walker, 2009). Public concern over radioactive waste skyrocketed after a large and serious leak of radioactive waste was widely reported to have happened at the Hanford, WA, site in 1973. This event lead Ralph Nader and the Union of Concerned Scientists to denounce the AEC's handling of radioactive waste at their own facilities and seriously question their ability to properly regulate commercial facilities (Walker, 2009). This and other scandals and controversies lead to the Energy Reorganization Act of 1974 that abolished the AEC and created the Nuclear Regulatory Commission (NRC). The same year the NRC created a task force charged with investigating and explicitly acknowledging the technical, political, organizational, and temporal uncertainties surrounding radioactive waste, including full consideration of societal and institutional issues (Walker, 2009). Persistent public concern mounted during the 1970s, with site suitability studies (for radioactive waste repositories) generating mass protests in MI, NY, and Washington, DC. At this point the DOE (created by President Carter in 1977) begins to focus on NM, NV, and WA as states that had been "nuclear friendly" in the past (Walker, 2009).

In 1982, Congress finally acted on the radioactive waste problem by passing the Nuclear Waste Policy Act (NWPA). Among other things, the 1982 Act identified nine possible sites in six states for further study regarding their suitability for an underground geological repository to permanently house high-level radioactive waste. The passage of the 1982 NWPA was possible because of a compromise between legislators from the eastern and western parts of the country regarding the location of sites that would be considered for the development of a repository. The Act included a provision of "regional equity" requiring the federal government to establish two geologic repositories—one east of the Mississippi River and one west of the river. Additionally, the 1982 Act specified the containment of radioactivity was to be accomplished by the *geologic* features of the repository site, not by engineered barriers.

The Nuclear Waste Policy Act Amendments (NWPAA) of 1987 abandoned the concept of "regional equity" and limited future site selection for a radioactive waste repository to just one site: Yucca Mountain, NV, 90 miles north of Las Vegas. Yucca Mountain is located on the ancestral land of the Western Shoshone and Southern Paiute, as acknowledged by the federal government in the congressionally ratified 1863 Treaty of Ruby Valley. The "1987 Screw Nevada Bill" (as it came to be known) left the state of Nevada and local American Indian groups holding the bag for the nation's first and only high-level radioactive repository unless (and even once) fatal flaws were discovered that would disqualify the site (Walker, 2009; Van Gerven, 2014; MacFarline and Ewing, 2007; Vandenbosh and Vandenbosh, 2007; Kuletz, 1998; Jacob, 1990). Since 1987, hundreds of citizen and environmental groups, American Indian Tribes and organizations, and the state of Nevada itself have fought the Yucca Mountain Project at every step of its development. Corbin Harney, a spiritual leader of the Western Shoshone, wrote in 2000:

> Yucca Mountain, in the heart of the Western Shoshone Nation, is a place of deep spiritual significance to Shoshone and Pauite peoples. Despite this, the federal government plans to send there 98 percent of the radioactivity generated during the entire Nuclear Age. . . . The government has no right to use Yucca Mountain this way. Newe Sogobia—the land guaranteed the Western Shoshone Nation by treaty—includes Yucca Mountain. Even the mere study of the site is a violation of the treaty. The Shoshone people have made their wishes clear: they want the DOE off their land and their mountain restored to them. (Harney, 2000)

According to Beyond Nuclear,

> Yucca Mountain was singled out in the first place due to raw politics, not sound science. Yucca Mountain is an active earthquake zone, at risk of volcanic activity, and would have massively leaked any radioactive wastes buried there into the drinking water supply below, as well as the air above. (Beyond Nuclear, 2010)

According to NIRS, more than one hundred international and national groups and over seven hundred state/local groups opposed the YMP (NIRS,

2010b). Perhaps most significantly, Native American groups from around the Southwest, including Western Shoshone and Southern Paiute groups, organized a "No Nuclear Waste on Native Lands" campaign. This campaign was directed at changing the fact that

> Every single proposal to store high level nuclear waste in North America targets Native territories. Not only do these proposals represent immense environmental injustices toward Native peoples but the dumps, if authorized, will enable a dying nuclear industry to get some last breaths. . . . Targeting isolated and economically disenfranchised Indians is their one solution. Help us close this loophole. Join the movement to stop nuclear waste on Native lands and create the impetus for our society to move towards wind, solar and other renewable resources. (Harney, 2000)

After years of "site suitability studies" the DOE recommended Yucca Mountain be developed for a high-level nuclear waste repository in 2002. Then-president Bush and Congress agreed, and in 2008 the Bush administration DOE submitted a license application to the NRC for the construction of the repository, which began a review process that was expected to take up to four years. However, in February 2010 the Obama administration made two crucial moves to end the Yucca Mountain Project. The FY 2011 budget the White House sent to Congress declared that the plan to store high-level radioactive waste at the Yucca site was "not a workable option." The budget proposed eliminating funding for the project office in the fiscal year that began October 1, including all funding for the Office of Civilian Radioactive Waste Management, which managed the day-to-day operations of the project. Secondly, Energy Secretary Steven Chu announced the Energy Department would file a motion to "stay" all license proceedings before the NRC for thirty days, and then would move to withdraw the license application "with prejudice." Withdrawing the license "with prejudice" means the department could not resubmit the license at a later date, and a new act from Congress would be needed to revive the project.

These moves by the Obama administration were met with much excitement and fanfare from all quarters of the antinuclear movement. For many groups (especially American Indian, environmental, and citizen groups in Nevada) these developments were seen as the payoff for more than two decades of relentless opposition and struggle against the YMP. The most active and long-standing opponents of the YMP include the grassroots group of the Nevada Nuclear Waste Task Force, the State of Nevada Agency for Nuclear Projects, as well as Western Shoshone Indian Organizations such as the National Council, Defense Project, Shundahai Network, and bands such as the Timbisha Shoshone Tribe in Death Valley.

Without these and other groups' tireless, and often thankless, efforts for over two decades, the fight against the YMP would have been lost long ago (Beyond Nuclear, 2010). Particularly for American Indian opponents of the project, this was not just an environmental victory, but also an environmental *justice* victory.

On the other hand, antinuclear group's excitement was tempered by the curious language the Obama administration used to frame their decision to terminate the project. The FY 2011 budget request only noted the Yucca Mountain Project was "not a workable option." Subsequent comments by administration and DOE officials consistently referred to the "social and political" obstacles hampering the YMP, and claimed as a result an alternative strategy would best serve the public's interests. At no point did the administration or DOE reference the numerous technical and/or scientific shortcomings of the Yucca Mountain site, which many in the antinuclear movement felt should have disqualified Yucca Mountain from consideration. In fact, as far back as 1998, 219 national and state/local groups unsuccessfully petitioned the DOE to immediately disqualify the Yucca Mountain, Nevada site, and declare it unsuitable for further consideration as a high-level nuclear waste repository (NIRS, 1998). This petition was filed in response to studies and data that antinuclear groups felt should have triggered automatic disqualification of the site pursuant to the 1982 NWPA. The NWPA clearly establishes the "suitability" of a site is defined in the Site Suitability Guidelines, which state a site is disqualified at any time during the siting process if evidence shows any disqualifying condition exists. The petition claimed the DOE had within its possession evidence and data that support the immediate disqualification of Yucca Mountain.

More specifically, using chlorine-36 as a tracer (which is only present in water exposed to atmospheric testing of nuclear weapons beginning in the late 1940s–1950s), residues from rainwater less than fifty years old had been detected at the underground depth of the proposed repository. This significant discovery contradicted earlier models of rainwater flow rates of surface water to the underground water table. This coupled with the groundwater flow model indicated the site met the conditions for disqualification under the Hydrology Guideline, which states a groundwater travel time to the accessible environment of less than 1000 years shall be grounds for disqualification (NIRS, 1998). Therefore the petition concludes

> The proposed Yucca Mountain repository will likely result in significant amounts of radionuclides degrading the quality of off-site supplies of groundwater that are presently suitable for and used for human consumption and crop irrigation. . . . This concern violates . . . the Disqualifying Condition for Socioeconomic Impacts. (NIRS, 1998)

These deficiencies with the Yucca Mountain site along with well-documented concerns over seismic activity, volcanic activity, and human intrusion at the site lead the antinuclear movement to conclude Yucca Mountain could never be shown to be a suitable location for a high-level radioactive waste repository. However, none of these issues were raised by the Obama administration and/or the DOE as grounds for terminating the YMP in 2010. Instead, the decision to terminate the project was repeatedly framed as a result of unsurpassable "social and political obstacles" that resulted in the need for a new strategy regarding the storage of nuclear waste. From a political/organizational point of view, it is entirely understandable why the DOE would not reference the technical/scientific shortcomings of the Yucca Mountain site, as the DOE had been denying the significance (and even the existence) of these data for many years. It would have been essentially impossible for them to officially admit there were in fact technical deficiencies with the Yucca Mountain site that should have rendered it disqualified years ago. To do so would have thrown light on the widespread practice of "pencil engineering," whereby officials respond to the failure of a site to meet some preestablished criteria by changing or reducing the threshold for compliance with that criteria. This is exactly what the DOE did with regard to the data relating to water flow rates. When data clearly showed water moved from the surface into the repository and back out again much, much faster than previously assumed, the response was not to disqualify the site, as required by the 1982 NWPA, but rather to change the guideline.

The administration's and DOE's failure to reference any of these technically disqualifying features of the Yucca Mountain site opened the door for pronuclear politicians and organizations to claim the decision to terminate the project was entirely "politically motivated." More specifically, pronuclear groups such as the Nuclear Energy Institute and legislators from South Carolina and Washington State (among others) slammed the decision to terminate the Yucca Mountain Project as a "politically motivated decision" that reflected a backroom deal between President Obama and then-Senate Majority Leader Harry Reid (D-NV). Reid had opposed the YMP during his entire tenure in the Senate, having suffered the humiliation of the "screw Nevada bill," which limited site selection to only Yucca Mountain, during his freshman year. Critics of the project's termination claimed Obama had promised Reid he would end the project if Reid helped deliver the States of Nevada to Obama during the 2008 presidential election. The legitimacy of these claims is essentially impossible to determine. However, the consistent failure to reference *any* of the technical problems with Yucca Mountain surely provided fuel for these critics. For many organizations in the antinuclear movement, it appeared the Obama administration had made the right decision, although possibly for the wrong reasons.

Despite the ongoing legal battle over the termination of the Yucca Mountain Project within the NRC and District Courts, by the end of September 2010 work on the project was essentially terminated. Concurrent with the administration's announcement (February 2010) that the Yucca Mountain Project would be canceled, President Obama instructed Energy Secretary Stephen Chu to form a Blue Ribbon Commission on America's Nuclear Future (the Commission or BRC). The Executive Order addressed to the Energy Secretary dated January 29, 2010, begins by saying:

> Expanding our Nation's capacity to generate clean nuclear energy is crucial to our ability to combat climate change, enhance energy security, and increase economic prosperity. . . . An important part of a sound, comprehensive, and long-term domestic nuclear energy strategy is a well-considered policy for managing used nuclear fuel and other aspects of the back end of the nuclear fuel cycle. (Executive Order, 2010)

This introductory language and the use of the phrase "America's nuclear *future*" in the BRC's title indicate the commission's foundational assumption that the use of nuclear power in the United States would not only proceed in the future but would be expanded. The question of whether or not the United States should continue to use (or expand) nuclear power had clearly already been answered in the affirmative. Thus, the limited scope of the kinds of questions the BRC could ask and answer severely undercut all of the BRC's subsequent efforts to proceed openly and transparently. For the antinuclear movement *the most important question* (as to whether or not the United States should continue to produce nuclear power) was never open for public discussion, and thus the decision reached did not include public input. As will be discussed more below, this is the very definition of misrepresentation and/or "misframing," as laid out by Fraser (2010), in that the rules governing the decision-making process in this case preclude public participation in *the most fundamental* decision(s) regarding nuclear power.

After noting the importance of sound radioactive waste disposal policy, the Executive Order goes on to say, "Yet the Nation's approach, developed more than 20 years ago, to managing materials derived from nuclear activities, including nuclear fuel and nuclear waste, has not proven effective" (Executive Order, 2010). As noted above, the EO does not specify why past approaches to radioactive waste disposal were not effective, only that they were. Accordingly, the Executive Order instructed the Secretary of Energy to create the commission and appoint its members. Those members were to include recognized representatives and experts from a range of disciplines and with a range of perspectives. (Although, for many antinuclear groups, the range of perspectives included did not range far enough beyond those of the nuclear industry and their allies in government.) The commission's proceedings were to be "open and transparent." The commission was tasked with performing a comprehensive review of policies for managing the back end of the nuclear fuel cycle, including all alternatives for the storage, processing, and disposal of civilian and defense used nuclear fuel and nuclear waste. More specifically, the commission was instructed to consider a broad range of technological and policy alternatives, and to analyze the scientific, environmental, budgetary, economic, financial, and management issues, among others, surrounding each alternative it considers (Executive Order, 2010).

In keeping with the fundamental assumption guiding the commission (that the United States would continue or expand the use of nuclear power in the future), the BRC was instructed to only consider policies for managing "the back end of the nuclear fuel cycle." This is consistent with the broader pattern of hazardous waste regulation in the United States first identified by Szasz (1994) in his book *Ecopopulism: Toxic waste and the Movement for Environmental Justice*. In this work, Szasz argues the passage of the Resource Conservation and Recovery Act in 1976 settled two long-standing questions about the regulation of hazardous (including radioactive) materials in the United States. One question had to do with whether to regulate the production or regulate disposal. In other words, should the government regulate industrial waste by regulating production, and make industry reduce the amount of waste it generates, or should the government not interfere with the production side and instead focus on improving the quality of disposal? The second question had to do with who would construct, own, and operate hazardous waste facilities. As Szasz argues, regulation creates demand for hazardous waste disposal/handling facilities, not necessarily supply. So the question was, should the federal government step-up and take the lead in building and operating hazardous waste facilities, or should they be more passive and wait for private sector entrepreneurs to see the business opportunity?

The RCRA of 1976 answered both of these questions, at least partially and temporarily. It became established that the federal government would focus its regulatory efforts on the *disposal* of hazardous waste in the form of "tailpipe regulations," thus by and large not interfering in the productive affairs of industry. Additionally, it was established the federal government would (at least initially) take the lead in constructing and operating hazardous waste facilities until the private sector could be more fully developed (Szasz, 1994). This regulatory strategy is continued with regard to the regulation of high-level radioactive waste. The government decided to deal with the radioactive waste problem at the "tailpipe" or backend of the fuel chain, rather than on the production side of the equation. Also, as established in the 1954 Atomic Energy Act and the 1982 NWPA, and continued with the establishment of the BRC, the federal government would build, own, and operate one or more high-level radioactive waste disposal facilities.

However, this regulatory strategy often puts regulatory agencies at odds with movements for environmental justice, including the antinuclear movement. The explicit decision to focus regulatory efforts (and thus governmental efforts to protect people and the environment) on the "tailpipe" or backend of the production process means potential protective actions are not being taken at the frontend, where the hazardous materials are actually created. The processes that lead to the creation of hazardous materials are often just as (if not more) important as issues of disposal to environmental justice activists, including the environmental justice energy activists in the antinuclear movement. The result in many cases is dissatisfaction on the part of environmental justice (energy) activists with public involvement in decision-making procedures. It is very common for concerned citizens and social movement activists to feel they have played a small, easily dismissed role in environmental and technological decision-making procedures. This is often due to the fact that participants come to understand many of the most fundamental and impactful decisions have already been made, and thus are "outside the scope" of public involvement.

As it happens, this dynamic often seems to escape the attention of regulatory officials and representatives of polluting industries. Many people and groups, on the other side of the table from the public in these proceedings, often publicly express their confusion and bewilderment about why public participation programs are not considered more legitimate by members of the public. As a result, they tend to rely on more and more "educational" efforts and campaigns, as if the problem is a lack of information on the part of the concerned public (Freudenberg and Grambling, 1983, 2010). What they fail to understand in many cases is concerned members of the public are often exceedingly well informed about the issues at hand. Their concerns are not irrational nor based on ignorance; they are often very rational and based on the perceived injustice of limited democratic involvement in decisions that may drastically affect their lives.

Somewhat apart from the problematic regulatory paradigm under which the BRC was created and operated, the composition of the commission itself initially gave the antinuclear movement reason for limited optimism, as well as reason for serious concern. As mentioned earlier, by and large the fifteen-member commission consisted of former legislators and nuclear regulatory officials, representatives from the energy and nuclear industries, labor leaders, and other academic experts from the fields of public policy, environmental science, and nuclear physics, and engineering. Conspicuously absent from the commission were any representatives from communities impacted by the various stages of the nuclear fuel chain, or representatives from the antinuclear movement. According to Mary Olson of the Radioactive Waste Project at NIRS,

> a majority of the Commissioners are individuals who have made, or supported the making of, the radioactive waste in question over the course of their careers. . . . The Blue Ribbon Commission on America's Nuclear Future does include several members who are not directly tied to the nuclear industry, but a controlling share of the seats are held by individuals who, at one time or another, have had primary decision-making authority, or who have personally profited from commercial nuclear technology. (NIRS, 2012a)

This is a clear case of what Fraser (2010) calls the injustice of "ordinary misrepresentation," where some people impacted by the decisions reached do not have adequate representation in the decision-making process. In this case, the injustice of "ordinary" misrepresentation is inflicted on those people and groups who are perhaps *most* impacted by the decisions reached regarding the handling of high-level radioactive waste (e.g., American Indian organizations, uranium mining and nuclear reactor communities, etc.).

The commission was cochaired by *Lee Hamilton*, who helped lead the 9/11 Commission, and *Brent Scowcroft* who had experience as National Security Advisor to President Ford. For some antinuclear groups there was limited optimism that Hamilton would be sensitive to the dangers nuclear reactors and radioactive waste pose as terrorist targets, and Scowcroft would be wary of reprocessing technologies, which were banned during his time in DC. The other thirteen commissioners included the following. *Mark Ayers*, president, Building & Construction Trades Department, AFL-CIO. The AFL-CIO is explicitly pronuclear, reflecting their belief that large, centralized construction projects (e.g., new nuclear reactors, disposal facilities, etc.) are easier to unionize than smaller, more decentralized energy projects like those associated with renewable energies; *Vicky A. Bailey*, principal, Anderson Stratton Enterprises, LLC. Bailey had over twenty years of corporate executive and governmental experience in energy and regulated industries, including experience as a public utility corporate executive and as the leading international official for the U.S. DOE; *Albert Carnesale*, chancellor emeritus and professor of public policy and mechanical and aerospace engineering at UCLA. Carnesale was a member of the Mission Committees of the Lawrence Livermore National Laboratory and the Los Alamos National Laboratory; *Pete V. Domenici*, Senior Fellow, Bipartisan Policy Center and former U.S. Senator (R-NM). During his years in the Senate Domenici earned a reputation for being one of the most staunch proponents of nuclear power in Washington, and authored the book *A Brighter Tomorrow: Fulfilling the Promise of Nuclear Energy* in 2004; *Susan Eisenhower*, president,

Eisenhower Group, Inc., which provided strategic counsel on political, business and public affairs projects for Fortune 100 and Fortune 500 companies, including several nuclear firms and utilities. Eisenhower had experience as an advisor to the DOE on issues of nuclear-nonproliferation and the threats of nuclear terrorism; *Chuck Hagel*, former United States Senator (R-NE). Hagel served on the board of directors of Chevron Corporation, the advisory boards of Deutsche Bank Americas, Corsair Capital, M.I.C. Industries, was a Director of the Zurich Holding Company of America, and was a senior advisor to McCarthy Capital Corporation. Through his numerous connections with these financial organizations, Hagel had many economic, political, and social ties to the nuclear power industry; *Jonathan Lash*, president, World Resources Institute, which is a nonprofit environmental organization that conducts research and policy analysis intended to produce practical solutions to global sustainability challenges. Mr. Lash was a recognized expert on climate change, energy security, and environment and development policies, and could be considered one of two environmentalists on the commission; *Allison Macfarlane*, associate professor of Environmental Science and Policy, George Mason University and coeditor (Rodney C. Ewing) of *Uncertainty Underground: Yucca Mountain and the Nation's High-Level Nuclear Waste* (MIT Press), which seeks to interrogate the uncertainties found within the different technical dimensions of the Yucca Mountain Project. She was a Social Science Research Council-MacArthur Foundation fellow in International Peace and Security and had served on National Academy of Sciences panels on nuclear energy and nuclear weapons issues. She had also served as chair of the Science and Security Board of the Bulletin of the Atomic Scientists and served on the Keystone Center's Energy Board. Her research focuses on environmental policy and international security issues associated with nuclear energy, especially the back end of the nuclear fuel cycle. While she has previously supported the Yucca Mountain Project, Macfarlane had a much more realistic understanding of the pros and cons of nuclear power, than did most of the other commission members. For instance, in a 2003 article Macfarlane showed that in order for nuclear power to significantly reduce the amount of CO_2 produced in the United States it would require a tenfold increase in the number of nuclear reactors presently operating. This tenfold increase would result in the production of about 70,000 tons of high-level radioactive waste *per year*, which would require constructing and opening a geologic repository the size of Yucca Mountain *every year*!; *Richard A. Meserve*, former chairman, U.S. Nuclear Regulatory Commission under Presidents Clinton and then Bush, where he oversaw the nearly disastrous debacle at the Davis-Bessie plant in Ohio; *Ernie Moniz*, professor of physics, Massachusetts Institute of Technology. Dr. Moniz served as Under Secretary of the DOE from October 1997 until January 2001. In that role, he had programmatic oversight responsibility for the offices of Nuclear Energy and Civilian Radioactive Waste Management, among others; *Per Peterson*, professor and chair, Department of Nuclear Engineering, University of California-Berkeley. Dr. Peterson had published over one hundred articles on the subjects of advanced nuclear reactor designs, inertial fusion energy, high-level nuclear waste processing, and nuclear materials management; *John Rowe*, chairman and chief executive officer, Exelon Corporation. Rowe had been a chief executive in the utility industry since 1984 and had been with Exelon Corporation since its inception in 2000. Chicago-based Exelon operates the largest fleet of nuclear power plants in the United States and had many long-standing ties with the Obama administration. Rounding out Rowe's pronuclear credentials is the fact he served as the chairman of Nuclear Energy Institute, which is the central lobbying and PR arm of the nuclear industry; finally *Phil Sharp*, president, Resources for the Future and former Representative (R-IN). At the time Sharp sat on the board of directors of the Duke Energy Corporation, which operated several dozen nuclear reactors in the Southeastern United States, and is a member of the National Petroleum Council.

These lengthy descriptions of the commission member's professional biographies have been included because these biographies clearly show the representational bias evident in the make-up of the BRC. What Kuletz (1998) accurately termed the "nuclear-industrial-academic-complex" was well represented among the commission members. The majority of the commission's members were former pronuclear legislators, former nuclear regulators with many long-standing ties to the industry, nuclear industry executives, and academic experts committed to nuclear technologies. Only Lash and Macfarlane could accurately be considered "environmentalists." Furthermore, and perhaps most importantly, there were no representatives of communities who were negatively impacted by the nuclear fuel chain, nor any representatives from any quarter of the antinuclear movement. This is precisely the type of political injustice that Fraser (2010) refers to as "ordinary" misrepresentation, in that decision-making procedures (regarding the future of radioactive waste disposal) are set up in such a way as to exclude some people and groups who are affected by the decisions reached. As will be argued here, the misrepresentational aspects of the BRC's composition resulted in widespread dissatisfaction, within the antinuclear movement, with both the commission's proceedings and their eventual recommendations.

Overall, before the BRC even held their first public meeting, it was becoming increasingly clear that the commission was being built around representational injustices on at least two levels—what Fraser (2010) calls the political injustices of "misframing" and "ordinary misrepresentation." The former relates to the fact that *the central question* regarding whether or not to continue to pursue nuclear power was framed as being "outside the scope" of the BRC, and thus did not include public participation or even comment. The latter relates to the fact that the structure of the decision-making process functions wrongly to exclude those who (in this case) are most impacted by the decision reached regarding the handling of radioactive wastes (e.g., American Indian groups, reactor communities, uranium mining communities, etc.). These injustices worked in tandem to seriously undercut the perceived legitimacy of the commission itself, as well as the recommendations that resulted, for those in the antinuclear movement.

During the first public meeting of the full BRC on March 26, 2010, Kevin Kamps of Beyond Nuclear gave testimony urging the commission to "Put a stop, once and for all, to the shameful history of targeting Native American communities and lands with radioactive waste dumps." Kamps is the "radioactive waste watchdog" for Beyond Nuclear and specializes in issues of high-level waste management and transportation, new and existing reactors, reactor decommissioning, and federal subsidies related to nuclear technologies. In 2005, Kamps along with Public Citizen documented the history of "radioactive racism" at the heart of national efforts to place high-level radioactive waste in "parking lot dumps" on native lands (Kamps, 2005).

Briefly, "Radioactive Racism: The History of Targeting Native American Communities with High-Level Atomic Waste Dumps" expresses much of the antinuclear movement's concerns regarding the environmental injustices associated with radioactive waste disposal and nuclear energy production in the United States. With the passage of the NWPAA in 1987, Congress created the Office of the Nuclear Waste Negotiator in an effort to open a federal "Monitored Retrievable Storage" (MRS) site for the temporary, interim storage of high-level nuclear waste. The Negotiator was authorized to seek states, counties, or Native American Tribes that might be interested in hosting such a facility in return for compensation. The process was supposed to be "voluntary," where a state, county, or tribe's elected representatives could only act under the authorization of the majority of their people. However, in a mistake that would be repeated almost twenty-five years later by the BRC (see below), no procedures were established for determining what constituted a "voluntary" host community.

"Voluntary siting" has long been seen as a representatively more just form of hazardous facility siting than are more traditional methods (e.g., the use of "eminent domain"). However, especially when dealing with American Indian communities, "voluntary siting" often does not reflect nor promote more socially and environmentally just methods of decision-making.

> The economic impoverishment of tribes, the tendency for tribal leaders to act without the authorization or even knowledge of their people, and the Bureau of Indian Affairs' ability to interfere in internal tribal affairs arbitrarily and capriciously . . . all mean that these siting processes were all too often not fair, nor truly voluntary. (Kamps, 2005)

In other words, injustices of maldistribution (tribal impoverishment), misrepresentation (tribal governance), and misrecognition (BIA) work in tandem to prevent the "voluntary" siting of atomic waste on Native lands from being socially, politically, economically, and/or environmentally just.

In August of 1990, David Leroy was confirmed by Congress as the first Nuclear Waste Negotiator, and in May of 1991 his office sent letters to every state, county, and federally recognized tribe in the country, offering hundreds of thousands (and eventually millions) of dollars for first considering, and then ultimately hosting, a dump. Of the fifty states and thousands of counties approached, only four counties officially respond, which represents about a 0.1 percent response rate. Out of the over five hundred federally recognized tribes approached, twenty tribes applied for Phase I study grants, which was about a 4.0 percent response rate, or almost forty times higher than that of counties (Kamps, 2005). Clearly there was not great interest in "volunteering" for this type of facility on the part of political units on various levels, or on the part of tribal governments. By August 1992 the four counties that responded to the Negotiator's request were no longer moving forward, due to significant local opposition and the opposition of the State Governors. Consequently, at this point the Negotiator (and the entire siting process for an MRS facility) began focusing exclusively on Native American communities.

In fact, during a truly revealing speech Leroy gave in 1991 before the National Congress of American Indians (which is comprised of tribal chairpersons and is the oldest Native American organization in the United States) titled "Federalism on Your Terms: An Invitation for Dialogue, Government to Government," the Negotiator argued Native American Tribes are incredibly well suited (even preferable) for storing the country's high-level nuclear waste. Quoting the famous Duwamish leader Sealth (more commonly known as Chief Seattle) many times, Leroy cites the Native American values of long-term health and sustainability as reason for this belief (Kamps, 2005). More specifically, Leroy infamously argued:

> It is the Native American cultures of this continent which have long adhered to the concept of planning for many generations of future unborn children in the decisions which are made today. . . . With atomic facilities designed to safely hold radioactive materials with half-lives of thousands of years, it is the Native

> American culture and perspective that is best designed to correctly consider and balance the benefits and burdens of these proposals. (quoted in Kamps, 2005)

For many American Indian representatives in attendance, Leroy's connecting some ill-conceived idea of a "Native ethic" with nuclear waste storage was not only patronizing but also "the granddaddy of all oxymorons." Grace Thorpe, the renowned American Indian activist, bristled at the federal agency's attempt to co-opt tribal traditions in 1993 when speaking before the National Congress of American Indians by saying, "It is wrong to say that it is natural that we, as Native Americans, should accept radioactive waste on our lands, as the U.S. Department of Energy has said" (quoted in Kamps, 2005).

By August of 1993, only the Mescalero Apache Tribe (New Mexico) and the Skull Valley Band of Goshutes (Utah) were still moving forward with preliminary negotiations regarding an MRS site. In October of 1993, Congress voted to effectively end the work of the Nuclear Waste Negotiator and the office's funding expired in December of 1994. This was largely due to opposition from states like New Mexico who feared an MRS site would be sited without their permission, control, or profit (Kamps, 2005). Also, the nuclear power industry and its supporters in Congress had grown impatient with the lack of success of the Negotiator and instead began pushing for "interim storage" of high-level radioactive waste, a strategy that would be continued over two decades later by the BRC (see below). Additionally, grassroots Native American activists like Grace Thorpe also played a very important role in the demise of the Negotiator program (Kamps, 2005).

However, by December of 1993 thirty-three nuclear utilities organized themselves into a private consortium and picked up negotiations (where the Nuclear Waste Negotiator left-off) with the Mescalero and Skull Valley tribes. By March 1994 the consortium was in serious negotiations with the Mescalero Tribe, which involved the consortium sending large amounts of money to the Tribe's Council Members (Kamps, 2005). At this point, the lack of democratic decision-making (and thus representational justice) became blatantly apparent, as did the general problems with seeking "volunteer" host communities without specifying how these would be determined. Rufina Marie Laws, a Mescalero Apache living on the reservation, began publically opposing the dump and formed a group called Humans Against Nuclear-Waste Dumping (HANDS) (Kamps, 2005). By the end of 1994, the consortium had reached a tentative agreement with Mescalero Tribal Council. Although the Tribal Council had been in negotiations resulting in this agreement for over three years, tribal members themselves knew little about the proposal. No public meetings had been held, despite a petition being signed by 221 tribal members in 1992 requesting such a meeting to discuss the pros and cons of the proposed facility. Although the Tribal Council refused the petition's request (to hold a public meeting on the issue), tribal members invited speakers and held the meeting anyway in August of that year (Kamps, 2005). When the proposal to host the MRS dump finally come before the Tribe for a vote in January of 1995, the Mescalero Apaches voted 490 to 362 to deny the proposal.

But in March 1995 a second petition drive calling for a second referendum was under way. Although tribal officials characterize the second petition drive as a "grassroots initiative," the move to overturn the referendum was led by the Tribal Housing Director. This proved to be especially problematic because as the tribal official heading up the petition drive was also in charge of tribal housing and other support services, many tribal members feared voicing opposition to the dump, lest they suffer retaliation and loss of services (Kamps, 2005). Although the signature sheets were never made public, it was determined by tribal officials that the petition had gathered enough signatures and a second vote took place. This vote overturned the previous referendum by a vote of 593 to 372, approving the dump on their land. Ironically, however, a month after Mescalero Apaches voted to "approve" the dump, the consortium of nuclear utilities dissolved due to the dissatisfaction of many of its thirty-three members with the progress it had made. In May of the following year, the Mescalero Tribe broke off negotiations with the consortium following continuous resistance and opposition by tribal members. At this point the original consortium reorganized itself with eight remaining members in Private Fuel Storage LLC (PFS) and resumed negotiations with the Skull Valley Band of Goshutes; the other Indian tribe to have proceeded past the first steps with Nuclear Waste Negotiator (Kamps, 2005).

The legitimacy of democratic/representative governance of the Skull Valley Band of Goshutes is similarly suspect, especially with regard to the issues of an MRS site for radioactive waste on their reservation. The push for such a facility began in 1992 when Lawrence and Richard Bear, chair and vice-chair of the Skull Valley Goshute Tribe, begin to make decisions without the knowledge of the General Council (the General Council refers to the entire adult membership of the Skull Valley Band). The Bears hired a "Tribal attorney" named Danny Quintana, a non-Indian who first brought the possibility of a waste storage facility to the Bears' attention, and who stood to capture a large share of the money the project would eventually bring. The Bears then created the unauthorized Tribal office of Project Manager (and appointed Leon Bear as the head), which proceeded to get involved in the U.S. Office of the Nuclear Waste Negotiator's MRS Project using tribal funds without informing or getting the approval of the General Council. At this point a Phase I grant of $100,000 from the Office of the Nuclear Waste Negotiator was awarded to the tribe to investigate the benefits and impacts of siting an MRS at Skull Valley, despite the fact the General Council had not approved involvement in this project (Kamps, 2005).

Dissatisfaction with the Bears' leadership continued to grow among tribal members, due in large part to the Bears' support for MRS dump, and on January 8, 1994, a recall election was held. Tribal political procedures required a quorum of the General Council. A quorum requires that more than half of the tribe's adult members attend the meeting. With forty-three of the seventy members of the General Council present, the Bears were voted out of office by a majority of the voting members present (Kamps, 2005). Additionally, a resolution was passed that rejected Danny Quintana as "Tribal attorney" and furthermore asserted that he never legitimately represented the tribe in any way whatsoever. Three days later, there was a meeting at the Headquarters of the Bureau of Indian Affairs in Salt Lake City, UT, concerning the recall elections. Despite acknowledging the recall elections were technically valid, the BIA decided it would not certify the election until the BIA had polled the entire General Council. This was due in part to Danny Quintana arguing all voting members of the General Council had not properly notified of the recall meeting/vote. The BIA agreed, although there is no legal basis in BIA procedure for this interference in an internal Tribal election (Kamps, 2005). On January 31, the BIA found their poll to be a tie between those who wanted a new election and those who were content with the recall proceedings. The BIA then ruled "the status quo should continue," which by their own definition would mean the recall elections stand (quoted in Kamps, 2005). However, the BIA determined that Lawrence, Richard, and Leon Bear would still be the Tribal officers. It's relevant to note that when asked two months later to produce the mailed-in poll ballots, the mailing list, the list of those returned, and so on, BIA says that it is unable to do so because all of the documents have been destroyed (Kamps, 2005).

This clear violation of Tribal sovereignty, and thus representative justice, set the stage for the eventual construction and operation of an MRS radioactive waste site on the Skull Valley Goshutes Reservation. In December of 1996, again without the knowledge or authorization of the General Council, Leon Bear signed a preliminary lease agreement with Private Fuel Storage, LLC (PFS) for an undisclosed amount of money. In May of 1997, an official lease agreement is signed by Leon Bear and PFS, again without the General Council's approval. The BIA approved the official lease agreement three days later, although the BIA claims to have no record of the process used to review the lease. The rubber-stamped approval of such an important document violates the BIA's trust responsibility to the Skull Valley Goshute Tribe and its individual members (Kamps, 2005). The final hurdle for PFS was cleared in May of 2000 when they signed a deal with Toole County (the county in Utah surrounding the Skull Valley Goshute Reservation) in exchange for the county's support of the dump. The agreement purportedly provided the county at least $90 million, and possibly more than $200 million, over the supposedly maximum 40 years that irradiated nuclear fuel could be stored on the reservation (Kamps, 2005). As part of the deal, the Toole County commissioners who were signatories to that contract were required not to say anything negative about PFS that might restrict or hinder PFS from moving forward in their license with the NRC. On February 24, 2005, the NRC's Atomic Safety Licensing Board approved PFS' preliminary license, and the full five-member NRC approved the official license in April of that year.

In his testimony before the Blue Ribbon Commission's first public meeting in 2010 Kamps sought to have this shameful, illegal, and unjust history of discriminatory siting policies explicitly *recognized* by the BRC. More specifically, Kamps repeatedly implored the commission to not only acknowledge how these unjust, discriminatory policies were in and of themselves wrong and indefensible, but also significantly contributed to the failures of previous efforts to craft nuclear waste disposal policy. Thus this history needs to be taken into consideration when thinking about how to proceed in new directions with regard to radioactive waste disposal. Moreover, he asked the commission to end this pattern of radioactive racism expressed in the targeting of Native Lands and Peoples for radioactive contamination. Sadly for Native American communities and the broader antinuclear movement concerned with energy justice, Kamps' pleas fell on deaf ears. The BRC essentially disregarded the injustices of these past efforts to develop radioactive waste disposal policy and recommended that moving forward the United States should pursue the same basic strategy of coercing impoverished, marginalized communities to "volunteer" to host such facilities. This glaring neglect of past and potential future environmental (in)justice issues by the BRC was succinctly summarized by the Blue Ridge Environmental Defense League (BREDL) in their response to the commission's eventual recommendations. BREDL stated in their response,

> [we] can find no reference for or commitment to the principles of environmental justice by the Blue Ribbon Commission . . . The Blue Ribbon Commission must not overlook important issues of environmental justice . . . [because] the ongoing struggle for environmental justice is part of the great change which seeks to alter society to become more humane towards respect for all living things. (BREDL, 2011)

As discussed above, the Blue Ribbon Commission on America's Nuclear Fuel was tasked with answering several "key questions" regarding the management, storage, and disposal of the nation's civilian radioactive materials, including the 70,000+ tons of high-level radioactive waste produced by nuclear power plants around the country. The BRC's Report to the Secretary of Energy released on July 29, 2011, contained seven "key" recommendations, as well as several suggestions for proposed legislative changes and tips for getting started with the process of siting new nuclear waste management facilities. Echoing the perspective of the BRC's Charter (discussed above) the report begins by noting,

> America's nuclear waste management program is at an impasse . . . The approach laid out under the 1987 Amendments to the Nuclear Waste Policy Act—which tied the entire U.S. high-level waste management program to the fate of the Yucca Mountain site—has not worked to produce a timely solution for dealing with the nation's most dangerous nuclear materials. (BRC, 2011, p. iii)

The report argues that the multiple reactor accident at Fukushima Daichii in March 2011 brought the problems of nuclear waste into the public eye as never before. Therefore, the report continues, a new strategy is needed for managing the back end of the nuclear fuel chain because continued failure to do so has already proven damaging and costly. They explain by damaging they mean damaging to the prospects of continued nuclear energy production; damaging to federal-state relations; damaging to public confidence in the federal government; and damaging to America's standing in the world as "a leader on global issues of nuclear safety, non-proliferation, and security" (BRC, 2011, p. iii). The costs of radioactive waste policy failure are said to fall on utility ratepayers, communities who have become unwilling hosts of long-term nuclear waste storage facilities (at existing reactor sites where the wastes were generated), and U.S. taxpayers who face mounting liabilities because of the failure of the federal government to meet waste management requirements.

Interestingly, the report's introduction also notes that beyond addressing these damages and costs, a new strategy is also needed for *ethical* reasons. The BRC states the "[current] generation has a fundamental ethical obligation to avoid burdening future generations with the entire task of finding a safe permanent solution for managing hazardous nuclear materials they had no part in creating" (BRC, 2011, p. iii). From this it is clear that the BRC was conscious of issues of fairness and social justice associated with radioactive waste policy, at least with regard to future generations. However, as discussed below, according to antinuclear SMOs the BRC's awareness of issues of social justice did not extend to considerations of current or historical inequalities between social groups. Thus antinuclear SMOs fear the general strategy the BRC recommended will continue the historical pattern of targeting marginalized communities (especially American Indian communities) for the development of nuclear waste storage sites, a practice that is rooted in and made possible by the existing distributive, representative, and recognitional injustices. In the remainder of this chapter, the BRC's recommendations are briefly reviewed before the responses of antinuclear social movement organizations are analyzed. Discourses of distribution, recognition, and representation structure the movement's various framings of the issue, which reveals the multiple and interrelated dimensions of the perceived injustices associated with the production of nuclear energy and radioactive waste storage. Furthermore, clear understandings of the perceived injustices will enable the development of a robust theory and application of principles of energy justice.

The strategy the BRC recommended had seven "key elements," the first of which was to develop a new, consent-based approach to siting future nuclear waste management facilities (BRC, 2011). Finding sites where all affected units of government (host state or tribe, regional and local authorities, and host communities) are willing to support or at least accept a facility has proven to be especially difficult. The crux of the problem stems from the federal/state/tribal/local rights dilemma, which is obviously not unique to the radioactive waste issue. However, the BRC argued that experience in the United States and around the world shows that attempts to force a top-down, federally mandated solution will take longer, cost more, and have lower odds of eventual success. Thus the approach they recommended is "explicitly adaptive, staged, and consent-based," which they believed could provide flexibility and sustain the public trust and confidence needed to complete such controversial facilities (BRC, 2011, p. vi). More specifically, the first step in siting new nuclear waste management facilities should be to develop a set of basic initial siting criteria, in order to ensure that time is not wasted investigating sites that are clearly unsuitable. The next step is then to develop a "generic standard" and to support regulatory requirements early in the siting process. The BRC believed preestablished and generally applicable standards and guidelines would be more likely to earn public confidence than site-specific standards, and would support the consideration of multiple sites (BRC, 2011). This included establishing initial program milestones laid out in a mission plan to allow for review by Congress, the administration, and stakeholders. The final and surely most controversial step was to "encourage expressions of interest" from a variety of communities that have potentially suitable sites.

"In practical terms, this means encouraging communities to volunteer to be considered to host a new nuclear waste management facility while also allowing for the waste management organization to approach communities that it believes can meet the siting requirements" (BRC, 2011, p. vii). Unsurprisingly, considering the history of radioactive racism discussed above, the antinuclear power movement expressed deep reservations about how "volunteer host communities" would be identified and selected. The BRC acknowledged the approach they recommended would involve lengthy negotiations with potentially affected state, tribal, and local governments and other entities. And in this context, they argued, any negotiation or selection process that was prescribed in detail up front is unlikely to work. Rather, "Transparency, flexibility, patience, responsiveness, and a heavy emphasis on consultation and cooperation will all be necessary" (BRC, 2011, p. vii). As analyzed in more detail below, the lack of specificity with regard to *how* "volunteer host communities" will be determined, identified, and selected was seen as problematic for antinuclear organizations considering the context of past policy failures. Despite the BRC's assurances and repeated usages of words like "transparency," "consultation," and "cooperation," many antinuclear power organizations failed to see how this new strategy was meaningfully different from the failed strategies of the past.

The second key recommendation contained in the BRC's report was to establish a new organization to implement the waste management program. While acknowledging some "notable successes" of the DOE and its predecessor agencies, the overall record of the DOE and the federal government with regard to nuclear waste does not inspire confidence.

> For this and other reasons, the Commission concludes that new institutional leadership is needed. Specifically, we believe a single-purpose, Congressionally-chartered Federal Corporation is best suited to provide the stability, focus, and credibility needed to get the waste program back on track. (BRC, 2011, p. viii)

The central task of the new organization would be to site, license, build, and operate facilities for the consolidated interim storage and final disposal of high-level nuclear waste. They recommended the organization be directed by a board nominated by the president, confirmed by the Senate, and selected to represent a range of expertise and perspectives. The third key recommendation (closely connected to the second) was for this new organization to have access to "utility waste disposal fees" and for these monies to be used for their intended purpose. The 1982 NWPA created a "polluter pays" funding mechanism to (theoretically) ensure the costs of disposing of commercial nuclear waste would be paid by utilities and their ratepayers, with no impact on taxpayers and the federal budget. A fee was assessed on every kilowatt-hour of nuclear-generated electricity as quid pro quo payment in exchange for the federal government's legal obligation to begin accepting commercial nuclear waste beginning in 1998 (a commitment established in the 1982 NWPA). However, the fund has never worked as intended. A series of Executive Branch and Congressional actions have made the approximately

$750 million per year in annual fee revenues, and the unspent $25 billion balance at the time, effectively unavailable to the waste program (BRC, 2011). The commission noted that new legislation would be needed in order to transfer the unspent balance in the fund to the new waste management organization, and this would need to be done as soon as possible because it would enable key subsequent actions the commission recommends.

For instance, the fourth key recommendation was to begin prompt efforts to develop a new permanent geologic disposal facility.

> The conclusion that disposal is needed and that deep geologic disposal is the scientifically preferred approach has been reached by every expert panel that has looked at the issue and by every other country that is pursuing a nuclear waste management program. (BRC, 2011, p. ix)

The BRC argued that deep geologic disposal capacity is an essential component of any comprehensive nuclear waste management system because very long-term isolation from the environment is the *only* responsible way to handle these materials. The commission explicitly did not take a position on the Obama administration's decision to withdraw the license application for the Yucca Mountain repository. However, they did note that regardless of what happens with Yucca Mountain the country's inventory of nuclear waste would soon surpass the total volume that could be stored there. Therefore, the United States would have to find a second disposal site even if Yucca Mountain went forward, and they believed the approach they set forth provided the best strategy for assuring progress (BRC, 2011, p. x).

The fifth key recommendation made by the BRC was to begin prompt efforts to develop one or more "consolidated interim storage" (CIS) facilities. This was one of the most problematic recommendations from the point of view of the antinuclear power movement. The BRC argued that interim storage is another crucial element of a national waste management system because it would allow the federal government to begin the orderly transfer of nuclear waste from reactor sites to "safe and secure" centralized facilities independent of the schedule for operating a permanent repository (BRC, 2011, p. x). The arguments in favor of consolidated storage are strongest for "stranded" waste located at shutdown plant sites, and this waste should be the first in line to be transferred so the sites can be completely decommissioned. Furthermore, the commission argued consolidated storage would provide valuable "flexibility" in the waste management system that could provide significant cost savings for both ratepayers and taxpayers when a significant number of plants are shut down in the future. To allay state and community concerns that a temporary storage site could become a de facto permanent disposal site, the commission argued the program to establish a consolidated storage site needs to be accompanied by a parallel disposal program that was "effective, focused, and making discernible progress in the eyes of key stakeholders and the public" (BRC, 2011, p. x).

The commission noted they received extensive public comments that showed many people fear the transportation of nuclear waste that will be a part of any consolidated storage or disposal program. Tellingly, this is the only explicate mention of public comments contained within the report. Once a temporary or permanent site has been developed the 70,000+ tons of high-level radioactive waste will have to be moved from the seventy reactor sites where it was produced and is currently stored via trains, trucks, and barges. Despite noting the extensive public comments of concern, the commission argued the existing standards and regulations for transporting radioactive waste had "functioned well and the safety record for these types of materials has been excellent" (BRC, 2011, p. xi). Nevertheless, the commission argued state, local, and tribal officials must be involved in efforts to communicate with the public and address these concerns, and so must be given the information and resources needed to carry out these roles and obligations.

The sixth and seventh recommendations made by the BRC were for the United States to support advances in nuclear energy technology and to take an active leadership role in international efforts to address safety, security, and non-proliferation concerns. Unsurprisingly given the composition of the commission (discussed above) they claimed "advances in nuclear energy technology have the potential to deliver an array of benefits across a wide range of energy policy goals" (BRC, 2011, p. xi). They recommended the NRC develop a regulatory framework for "advanced" nuclear energy systems that would lower barriers to commercial investment by increasing confidence that new systems could be successfully licensed. Additionally, as more nations (potentially) begin producing nuclear power the United States urgently needed to take on a leadership role on issues of safety, non-proliferation, and security. However, according to the commission, the United States could not exercise effective leadership on these issues related to the back end of the nuclear fuel chain as long as its own program was in disarray; effective domestic policies were needed to support the U.S. international agenda.

The report concludes by again noting that the Fukushima disaster had focused new attention on issues of nuclear safety around the world. The commission felt the nuclear waste problem was somewhat unique in that there was wide agreement (or so they claimed) about the "outlines" of a solution. The commission argued that we knew what we need to do, we knew we had to do it, and we even knew how to do it. The core difficulty was the same as it had been for thirty years:

> finding a way to site these inherently controversial facilities . . . in a manner that allows all stakeholders, but most especially host communities, states, and tribes, to conclude that their interests have been adequately protected and their well-being enhanced—not merely sacrificed or overridden by the interests of the country as a whole. (BRC, 2011, p. xiv)

The commission believed the approach they suggested, with the emphasis on words like "transparency," "flexibility," "accountability," and "meaningful consultation," was the most likely to produce positive results moving forward.

However, for the antinuclear movement simply repeatedly using these legitimacy-granting words and phrases was not the same thing as actually developing an *alternative* technically/environmentally defensible and socially/politically just plan for dealing with the radioactive waste crisis. That said, the antinuclear movement supported some of the commission's recommendations, but overall the movement's responses reflected the movement's perception that there was far more the same than different between what the BRC recommended and the failed policy approaches of the past. From the movement's perspective, not only did the BRC not address past representative, recognitional, and (potential) distributive injustices associated with nuclear waste policy, but they also left the door open for a continuation of these multidimensional injustices moving forward. The failures to address these interrelated injustices were widely recognized to have contributed to the past failures of radioactive waste disposal policy (although not necessarily in these terms). Consequently, the antinuclear movement's responses to the BRC's recommendations drew from and reconstructed the social justice paradigms of distribution, recognition, and representation. The movement's agreements and disagreements with the commission's recommendations were founded upon these interrelated dimensions of energy justice. Therefore, an analysis of the antinuclear movement's responses to the commission's recommendations facilitates the identification of additional principles of just energy systems, which can help guide future policy to ensure social, political, and economic justice is associated with future energy systems.

Immediately following the release of the BRC's report Beyond Nuclear released a statement expressing their general disappointment with the commission's recommendations. They began by noting, "There remains no viable solution for either the management or certainly the 'disposal' of radioactive waste. Yet, the one essential recommendation that is not contained in the DOE report is to stop making any more of it" (Beyond Nuclear, 2012b). With regard to high-level radioactive waste, the antinuclear movement's first and most-insistent demand is to "stop making more of it!" Given the past and (likely) future difficulties creating technically/environmentally sound and socially and politically/economically just radioactive waste disposal policy, the antinuclear movement feels it is extremely reckless to continue producing more of these materials without an adequate solution for their disposal. NIRS made the same argument in a list of talking points they put together to help people make public comments on the BRC's report:

> *Zero Point:* We support the identification of responsible radioactive waste management plans for the waste that currently exists. However we know, unequivocally, that there is zero *guarantee* that the tens of thousands of tons of deadly material . . . will be contained and isolated from the biosphere for the hundreds of millennia that it will remain a hazard. Therefore it is only logical, based on the principle of *precaution*, that [the] report clearly reflect that *stopping the generation of any additional radioactive waste* will assist the implementation of any reasonable and responsible waste management plan. (NIRS, 2012b, emphasis in original)

In fact, this basic point about the irresponsibility and injustice of producing such hazardous substances was also made by Grace Thorpe back in 1993 during a speech before the National Congress of American Indians. Thorpe leaped over the normal foci of technocratic driven public debates about radioactive waste disposal (such as siting criteria, program milestones, millirems, etc.) to pose the central question at the heart of these debates; "What kind of society permits the manufacture of products that cannot be safely disposed of? Shouldn't we have a basic law of the land that prohibits the production of anything we cannot safely dispose?" (quoted in Kamps, 2005).

Furthermore, the fact that the possibility of ceasing to produce more radioactive waste (thus ceasing to produce more nuclear-powered electricity) is not even considered anywhere in the commission's report is in part explained by the representational injustices associated with the commission's mandate and composition. As discussed above, from the Executive Order instructing the Energy Secretary to create the BRC, and the commission's Charter, it is clear that the foundational assumption guiding the BRC was the United States would continue to produce nuclear-powered electricity and high-level radioactive waste in the future. Thus, the most important question for the antinuclear movement (as to whether or not the United States should continue to produce nuclear power) was never open for public discussion, and thus the decision reached did not include public input. This is unsurprising given the composition of the commission itself, with the overrepresentation of interests that benefit from nuclear power, and the underrepresentation of interests that are burdened. This is a paradigmatic example of what Fraser (2010) refers to as the representational injustice of "misframing," in that the rules governing the decision-making process in this case precluded public participation in the most fundamental decision(s) regarding nuclear power. The question of whether or not the United States should continue to use (or expand) nuclear power had clearly already been answered in the affirmative. This limited the scope of the kinds of questions the BRC could ask and answer, and from the antinuclear movement's point of view, severely undercut all of the BRC's subsequent efforts to proceed openly and transparently. Because the commission failed to adopt "stop making waste" as its central theme, the movement argued the commission's "strategy" was really just a plan to enable the further production of *more* waste. Following from the movement's claims in this regard, an important principle of energy justice requires that energy systems be developed in such a way as to minimize the production of hazardous by-products associated with energy production to the greatest extent possible. This would work to minimize the socially and environmentally harmful consequences of energy systems, which have historically disproportionately burdened minority and low-income communities.

The link between hazardous material production and disproportionate burdens on low-income and minority communities, and thus the rational for this principle of energy justice, is supported by the BREDL response to the commission's recommendations. More specifically BREDL argues,

> The Blue Ribbon Commission must include steps to avoid disproportionate, adverse environmental impacts on low-income and minority populations and impacts of important religious, subsistence, or social practices. Finally, it should address the question of ending the production of dangerous materials which cannot safely be disposed. (BREDL, 2011)

BREDL defines environmental justice as:

> Environmental Justice is about social transformation directed towards meeting basic human needs and enhancing our quality of life—economic quality, health care, housing, human rights, environmental protection, and democracy. In linking environmental and social justice issues the environmental justice approach seeks to challenge the abuse of power which results in poor people having to suffer the effects of environmental damage caused by the greed of others. (BREDL, 2011)

Drawing from Schlosberg (2007) and expanding on this definition, BREDL explains:

> The principle of environmental justice incorporates 1 the equitable *distribution* of environmental risks and benefits; 2 the meaningful *participation* in environmental decision-making; 3 the *recognition* of community life, local knowledge,

> and cultural difference; and 4 the capability of communities and individuals to function in society. It means avoiding disproportionate adverse environmental impacts on low income populations and minority communities. (BREDL, 2011, emphasis added)

From this definition and discussion of environmental justice by a large, regional antinuclear organization we can clearly see the applicability of the environmental justice framework to the conflicts over the production of nuclear power and radioactive waste. More specifically, these claims by BREDL show that the conceptualization of environmental justice as consisting of interrelated dimensions of distribution, recognition, and representation is explicitly present within the antinuclear power movement itself. While it is somewhat uncommon for most antinuclear organizations to express their claims, desires, and grievances in precisely these terms (derived from the liberal political theory), these "families of claims" also operate as "folk paradigms of justice" that are often presupposed by social movement actors/organizations and consist of sets of linked assumptions about the causes of and remedies for injustice (Fraser, 2005). Most significantly, from this analysis and discussion, it is clear there is significant disagreement between the BRC and the antinuclear movement with regard to the environmental (in) justice implications of past, present, and future efforts to manage high-level radioactive waste.

In terms of agreement, the antinuclear movement generally agreed with the BRC's recommendation that one or more permanent, geological facilities for nuclear waste need to be developed. As noted in the report, virtually every expert assessment of the radioactive waste crisis has concluded that deep, permanent geological disposal is the best option for the possible containment of these materials from the biosphere for upwards of one million years. Many antinuclear organizations agreed with the commission that developing "generic" rather than "site-specific" standards and regulations, and doing so before site selection begins, will help build public confidence in a permanent waste repository program. As was seen with the Yucca Mountain Project, site-specific standards can too easily become "movable goalposts" in the service of advancing a particular site/decision, rather than protecting public health and the environment. The movement and the commission agree that having an established set of general guidelines and regulations will leave less room for the kinds of "pencil engineering" that characterized the Yucca Mountain Project, which was seen to have severely eroded public confidence in the program. As will be discussed in more detail in the concluding chapter, the development of generic standards for the protection of public health and the environment with regard to the implementation of energy systems is an important element in policy meant to achieve energy justice. Doing so helps to ensure the protection of public health and the environment remains central to decisions about energy production by working to ensure the best scientific knowledge and practices are put into the service of protecting social interests, rather than advancing primarily economic and/or political objectives.

Another area of agreement between the BRC and the antinuclear movement was the belief that the DOE had worn out its credibility and failed miserably to inspire confidence or trust in the U.S. nuclear waste management program. Many antinuclear organizations agreed that given the DOE's almost unblemished record of underperformance in this area, they should no longer be in charge of the waste disposal program. However, antinuclear movement organizations had serious reservations about, or even strongly opposed, the creation of a congressionally chartered federal corporation, and/or any move to privatize radioactive waste management. These reservations and concerns about the privatization of radioactive waste disposal programs were based upon the recognition of past representational and recognitional injustices, and the fear that these injustices would reemerge moving forward. For example, Physicians for Social Responsibility (PSR) argued their reservations about such an organization stem from the fact that other federal corporations, such as the Tennessee Valley Authority (TVA), had little congressional oversight, little public accountability, and often carry serious debt (PSR, 2011). NIRS argued that all aspects of radioactive waste management must be subject to the requirements of the Sunshine laws including the Freedom of Information Act and others that are meant to ensure transparency and public access to participation and information (NIRS, 2012b).

The basic concerns from the antinuclear movement's point of view were that the creation of a federal corporation to manage the radioactive waste storage program would further diminish and marginalize the already easily dismissed role public/community concerns play in the program. In other words, the movement feared the creation of a new federal corporation would further the representational injustices that favor the nuclear industry over the public, which leads to a misrecognition of public concerns, and ultimately a maldistribution of the burdens associated with radioactive waste disposal.

With regard to the antinuclear movement's disagreements with the BRC's recommendations, two issues were paramount: the logic of "centralized interim storage," and related concerns over the process for identifying/establishing "volunteer host communities." The antinuclear power movement in the United States at both the national and local levels has forcefully and consistently opposed all efforts to move radioactive waste from the site of its generation *before* a permanent disposal site is available. Be it in response to proposals to establish "MRS" in the 1990s or "centralized interim storage" today, the movement has argued that interim storage should be at the point of generation, and these extremely hazardous wastes should be moved only once. Reactor sites are already de facto interim storage sites, and the exact same storage technologies would be used at the new "interim" facility while incurring the additional risks of at least one extra round of transportation over the nation's roads, rails, and waterways. As NIRS has argued, "An 'interim site' is never the consolidation of waste unless and until the generation of new waste at all other sites stops. Until then, it is only one more site" (NIRS, 2012b). Like Superfund sites, NIRS argues the responsibility for the management of nuclear waste should extend to all corporations that contributed to and profited from the waste's generation.

Furthermore, PSR highlights the fact that the commission's recommendation that the "United States should proceed promptly to develop one or more consolidated interim storage facilities" ignores past failed efforts to develop such sites and fails to explain how such an effort will succeed *this* time around (PSR, 2011). The commission's explicate assumption that finding an interim storage facility "should be less difficult, particularly if it is accompanied by attractive incentives" did not comport with the past failures of the Nuclear Waste Negotiator (discussed above). The commission did not examine the relevant political, social, or economic reasons for these past failures, nor did they explain why compensation failed in the past, but was believed to work this time. Additionally, PSR brought to light how the reasons presented in the BRC's Report for recommending off-site "interim" storage were flawed. For example, according to PSR, the commission overemphasized the scope of the nuclear waste problem from so-called orphaned (permanently closed) reactor sites. At the time there are only ten of these sites in the country, and they could be addressed relatively easily on a case-by-case basis, rather than using them as the foundation for a decision to move the waste from more than seventy locations around the country.

> One or more "interim" sites will take at least a decade to find and license and another couple decades to package and transport the waste. This is not a small project and, as a result, will take attention and money away from the ultimate goal of developing a permanent geologic repository. (PSR, 2011)

In their report, the BRC claimed their desire to establish CIS was in part motivated by the commission's desire to be fair and just with communities that live adjacent to "orphaned" nuclear waste at closed nuclear reactor sites. However, according to a letter initiated by organizations representing communities around these permanently closed reactor sites, this recommendation runs exactly contrary to their desire to see radioactive waste completely isolated from the biosphere. The letter, addressed to then Secretary of Energy Stephen Chu, states:

> The Commission you appointed is claiming that it is acting in the interest of communities such as ours where closed nuclear power reactors are located, when in fact the Commission's recommendations are in opposition to our number one priority: isolation of radioactivity from our environment for as long as it is a hazard. Centralizing waste storage for purposes of expanded waste production or for reprocessing is contrary to this goal, and is not responsible policy. (NIRS, 2012a)

The commission was claiming to be working *for* these communities but was clearly not working *with* these communities. This is a clear-cut case of misrepresentation (lack of impacted community representation on the commission) leading to misrecognition of the desires and concerns of these communities. In all 88 national, regional, and local environmental organizations, more than 5,400 individuals signed this letter to Energy Secretary Steven Chu urging him to reject the recommendation from the BRC that would encourage the establishment of an "interim" radioactive waste storage dump and begin the transportation of high-level radioactive waste across the United States. Reflecting on the misrepresentation of these communities within the BRC's proceedings and the misrecognition of these community's concerns, Mary Olson of the Radioactive Waste Project of Nuclear Information and Resource Service said,

> Our voices have been heard, but disregarded. This comes as no surprise since a majority of the Commissioners are individuals who have made, or supported the making of, the radioactive waste in question over the course of their careers. Of course they want to move it—they want to make more. (NIRS, 2012a)

The letter initiated by organizations representing communities near closed reactors also undercut the BRC's argument that the lesions from Fukushima necessitated centralized storage. We now know the ongoing problems at Fukushima did not stem from the dry cask storage of nuclear waste at the site. The dry casks survived the earthquake and tsunami intact. Rather, it was the spent fuel pools that are vulnerable due to their continuous need for water and electricity. Therefore, it is unclear how the commission concluded a lesion from Fukushima is for the United States to move nuclear waste off-site. Doing so will not eliminate the need for spent fuel pools, as the fuel removed from reactors needs to be submerged in these pools for at least five years to cool. What needs to be done, according to this letter, is overcrowded fuel pools need to be reinforced and thinned out, with as much waste as possible moved to reinforced dry casks following the Principles of Hardened On-Site Storage (HOSS).

The Principles of HOSS have been signed by more than 170 national and local antinuclear organizations from all 50 states who agree HOSS is the most sensible option for addressing the *immediate* threats posed by nuclear waste at reactor sites. These principles are based on the belief that irradiated fuel must be stored as safely as possible as close to the site of generation as possible. Transporting waste away from the reactor site to an interim facility should only be done if the site of generation is unsuitable for HOSS (in which case the site should not have been licensed for a nuclear facility in the first place). HOSS facilities are not meant to be permanent, and thus should not be constructed deep underground. The waste needs to be retrievable and needs to be monitored with real-time radiation and heat measurements (NIRS, 2010c). The overall goal of HOSS is to make the amount of radioactive releases, even in the case of a severe attack, so low that the storage system would be an unattractive terrorist target. HOSS facilities need to be robust enough to withstand such things as a direct hit from a high explosive or a large plane full of fuel, which would also make them very effective against natural disasters such as earthquakes.

Given that even in the best-case scenario the United States is decades away from having operational "interim" storage or permanent disposal facilities. The antinuclear movement believes the United States needs to move as quickly as possible to secure the wastes, where they are currently located, to the greatest extent possible. In a letter sent by organizations representing reactor communities to the Energy Secretary, these groups asked the Secretary:

> Don't move the waste from the reactor sites for now—until a permanent isolation program is available. . . . Harden waste storage at the reactors—heed, adopt and implement the Principles for Safeguarding Nuclear Waste at Reactor Sites as has been repeatedly presented to the Blue Ribbon Commission. . . . Our sites are not suitable for waste "disposal" and so we will remain engaged until there is a path forward for permanent isolation of this waste for as long as it is a hazard. (NIRS, 2012c)

The fact that these requests were made by organizations representing the subject position of communities living with nuclear reactors gives these requests extra legitimacy and standing. The commission claimed they were working *for* the interests of these communities when they recommended centralized interim storage of nuclear waste. However, due to the lack of representation of these communalities within the commission, their interests were misrecognized and used as an excuse to further the interests of the nuclear industry at the expense of these communities' actual desires. The antinuclear movement exerts much of its efforts working to *oppose* various pronuclear policies and practices. The Principles for HOSS are one of the major policy proposals that the movement can strongly *support* and advocate for, so long as the implementation of HOSS is not used as an excuse to continue creating more radioactive waste. The fact that HOSS is strongly supported by groups representing reactor communities, who might appear to benefit from the moving of radioactive waste away from their communities to a centralized facility, takes away a big part of the legitimacy the commission was trying to wrap around their recommendation.

As discussed above, radioactive waste has long been seen as *the* crucial issue with regard to nuclear energy by both critics and proponents. Furthermore, the issues of radioactive waste storage and disposal are perhaps the most direct ways to see the environmental justice problems created by the production of nuclear power. The "unsurpassable social and political obstacles" that necessitated the canceling of the Yucca Mountain Project relate back to the representative, recognitional, and distributional injustices that have plagued the radioactive waste management program in the United States since the very beginning. Therefore, it is necessary to identify and correct these multidimensional injustices in order for the nuclear waste crisis to be solved in a technically, socially, and politically/economically acceptable manner.

The program for managing the 70,000+ tons of high-level radioactive waste produced in the United States will be unprecedented in human history. The eventual permanent disposal of these extremely hazardous materials will involve the construction of one or more facilities that must keep them isolated from all living things for upwards of one million years. For context, the Pyramids of Giza are roughly 5,000 years old; written language has existed for roughly 10,000 years; and *homo sapien sapiens* (modern human beings) have existed for roughly 200,000–300,000 years. The only nation to begin to address their high-level radioactive waste problem is Finland, which has begun the construction of a high-level radioactive waste repository. Although Finland's program is decades ahead of the program in the United States, and they have less than 10 percent of the total volume of radioactive waste in need of disposal, the project is not expected to be completed for at least another one hundred years. Therefore, even if the United States was ready to begin construction of a permanent disposal facility today, the project would likely require more than a century and extraordinary amounts of resources to complete. It follows from this (and the above analysis) that an important principle of energy justice requires that energy systems be developed in such a way as to minimize the production of hazardous by-products associated with energy production to the greatest extent possible. This would work to minimize the socially and environmentally harmful consequences of energy systems, which have historically disproportionately burdened minority and low-income communities.

Since the beginning of commercial nuclear power production the nuclear industry's and U.S. government's perception of the radioactive waste problem has evolved from "easily solvable" to "serious and growing" to "possibly without satisfactory solution." Public concern over the handling of radioactive waste in the United States began in the 1950s, and eventually resulted in the dismantling of the AEC and the creation of the NRC and later the DOE. Shortly thereafter, the 1982 NWPA began the current program to manage the nation's highly radioactive nuclear waste, and with the Amendments to this Act that were passed in 1987, the fate of the program was tied to the development of the Yucca Mountain repository.

That was until President Obama announced the cancelation of the Yucca Mountain Project in 2010 due to "unsurpassable social and political obstacles" that prevented the project from moving forward as planned. This was seen as a tremendous victory by the antinuclear movement in the United States, and especially as an environmental justice victory for American Indian groups living in the vicinity of Yucca Mountain. From the passage of the 1987 NWPA Amendments, which singled out Yucca Mountain as the only site to be considered for a permanent repository and created the Office of the Nuclear Waste Negotiator, the U.S. nuclear waste management program has been rife with representative, recognitional, and distributional injustices. As discussed above, the development of a nuclear waste repository at Yucca Mountain would have resulted in extensive environmental injustice for the Western Shoshone and Southern Paiute peoples who live in the region. In distributive terms, these minority, low-income groups would have been more negatively affected by the project than any other group in the general population. Moreover, American Indian groups in the Yucca Mountain region were not allowed to meaningfully participate in the decision-making process, and thus were subject to significant misrepresentation. The misrepresentation of American Indians in the decision-making process resulted in repeated misrecognition of their perspectives and concerns, and even the maldistribution of financial and other types of resources, beyond the maldistribution of the risks associated with the disposal of radioactive waste (Van Gerven, 2014).

The antinuclear movement's responses and reactions to the activities of the Nuclear Waste Negotiator in the 1990s show the multidimensional injustices of nuclear energy production and radioactive waste disposal are by no means limited to the Yucca Mountain Project alone. The Negotiator's efforts to find volunteer communities who were willing to host such facilities (in return for financial compensation) immediately and exclusively focused on impoverished American Indian communities. Thus the entire program was predicated upon taking advantage of preexisting distributive injustices that left these communities in situations desperate enough to consider the proposal. On top of this came the representative and recognitional injustices that stem from the somewhat unique social, political, and economic position of American Indian groups in the U.S. issues of Tribal governance, the democratic legitimacy of Tribal leaders, the influence of the Bureau of Indian Affairs (BIA), and others make questions of "who is volunteering whom for what" very difficult to answer. As discussed in more detail in the concluding chapter, the establishment of socially, politically, and economically just energy systems (i.e., energy justice) will require the explicit acknowledgment of preexisting injustices and inequalities, combined with concrete policy mechanisms to ensure the burdens are not being disproportionately bore by the most vulnerable.

As the above analysis has hopefully made clear, the injustices associated with efforts to manage nuclear waste are not just significant for their roles in past failures. Despite the initial excitement felt in the antinuclear movement about the cancelation of the Yucca Mountain Project, concerns quickly returned as the movement came to understand the mandate, assumptions, composition, and eventual recommendations of the Blue Ribbon Commission created to develop alternative strategies. From the charter and the mandate that created and guided the commission, deep, structural representational injustices were plainly evident. The decision as to whether or not the United States would continue to produce more radioactive waste (and nuclear power) had already been answered in the affirmative. The fact that what is seen by the antinuclear movement as *the* central question concerning radioactive waste management had already been answered without any public participation is a textbook case of what Fraser (2009) labels the representational injustice of "misframing." More specifically, misframing occurs whenever important aspects of the decision-making process take place beyond the scope of public input or participation, which is clearly the case here. Moreover, the composition of the commission itself reflects the other dimension of representative injustice, which Fraser (2009) labels "ordinary misrepresentation." While the interests, concerns, and perspectives of the nuclear-industrial-academic-complex (Kuletz, 1998) were well represented on the commission, there were no members representing impacted communities, environmental groups, or anyone who did not personally or professionally benefit from the continued production of radioactive waste and nuclear power.

Both dimensions of the representational injustices that structured the BRC severely undermined the legitimacy of the commission's claims to be taking public comment/participation seriously, as well as the commission's eventual recommendations. Despite this, however, there were some areas of agreement between the antinuclear movement and the BRC's recommendations. More specifically, the antinuclear movement generally agreed with the commission that the DOE had lost its credibility and should no longer be in charge of the radioactive waste management program, although the movement is extremely skeptical of any efforts to privatize the program in response to these failures. Additionally, there was agreement that the United States needs to establish one or more permanent, deep geologic disposal sites as the best long-term solution to the radioactive waste crisis. Significantly, there was also agreement that "generic standards/regulations" need to be established prior to the site selection process for such inherently controversial facilities. Both the commission and the movement agree that doing so will keep the protection of public health and the environment central in decisions about radioactive waste management, and thus will increase public confidence and acceptance of the program. As will be discussed in more detail in the concluding chapter, achieving energy justice through developing and implementing socially, economically, and politically just energy systems will require the creation of preexisting guidelines and standards meant to protect public health and the environment, which are less open to manipulation than site-specific guidelines. This principle of energy justice is *itself* consistent with the methodological framework established by Fraser (2000, 2003, 2009) as it was induced through an analysis of real-world claims for social justice, as opposed to preestablished abstract principles of justice. However, the *content* of this principle of energy justice requires the formation of preestablished guidelines and standards to protect public health and the environment. It is necessary to keep in mind the method of developing principles of (energy) justice is distinct from the content of the principles themselves, which in this case require preestablished, measurable rules and regulations.

Where disagreement between the commission's recommendations and the antinuclear movement was the greatest was around the issues of "centralized interim storage" and the prospects/procedures for establishing "volunteer host communities." These issues clearly reveal how representational injustices contribute to recognitional and (eventually) distributive injustices in this case. The commission claimed the need for centralized interim storage was largely driven by collective responsibility to communities living near permanently closed reactors that continue to store "orphaned" waste. However, there was no representative from these communities (or any other type of community impacted by nuclear power and radioactive waste) on the commission. The response from organizations representing these communities indicated the commission's recommendation misrecognized their desires and concerns. These community organizations and the broader antinuclear movement felt moving the waste before a permanent facility is in place, and before the production of more waste has ceased, was contrary to their desire for the waste to remain isolated from the biosphere. Speaking as the most proximate stakeholders, who could be seen as benefiting from the centralizing of waste storage, they want to see the waste made as safe and secure as possible where it is currently located, in order to avoid the creation of yet another nuclear site. The problems here stem back to the representational injustices that structured the commission's composition and procedures, which resulted in the misrecognition of these communities' concerns and desires.

The preceding analysis of the underlying discourses of justice utilized and constructed by the antinuclear movement in the United States with regard to radioactive waste management has revealed how issues of misrepresentation, misrecognition, and (potentially) maldistribution have continuously plagued the waste management program. For instance, the above analysis has highlighted the fact that energy justice requires energy systems to be developed in such a way as to minimize the production of hazardous by-products associated with energy production to the greatest extent possible. Also, the analysis has shown that energy justice would be advanced by developing and using "generic standards" to protect public health and the environment, which will help ensure these goals remain central to future decisions about energy production. Additionally, just energy policy needs to be based upon the explicit recognition of existing social and political/economic inequalities, and includes concrete mechanisms for ensuring the burdens of energy choices are not disproportionately borne by marginalized populations. These and other features of just energy policy that emerge from the analysis of the antinuclear movement's claims in other areas will be expanded upon and further theorized in the concluding chapter.

Chapter 3

Local Financing of New Nuclear Reactor Construction and Increased Nuclear Safety

As discussed in the introductory chapter, the politics of nuclear power production and the associated production of high-level radioactive waste in the United States take place on both the national and local levels. Consequently, in the United States the antinuclear power movement's opposition to the generation of nuclear power and radioactive waste is organized on both of these levels. Although much of the policy concerning the production of nuclear power and radioactive waste management is made at the federal level, it is the case that state and local level antinuclear organizations undertake localized campaigns to mitigate the socially and environmentally injurious aspects of this form of energy production. Local (as opposed to national) antinuclear organizations in the United States generally organize around, and focus upon, one or more specific aspects of nuclear power and radioactive waste production that impact and concern the communities/localities they represent.

These organizations are found in uranium mining communities where some local resident groups fight to have the contamination from past mining operations cleaned up and to prevent new uranium mines from opening. This opposition is also found in communities with uranium fuel fabrication and enrichment facilities, where some local stakeholder groups feel the environmental and human health costs of such facilities outweigh the economic benefits they provide. Grassroots opposition is also found in reactor communities where local stakeholder groups fight to shut down existing reactors, prevent existing reactors from being relicensed, and prevent new facilities from being licensed and constructed. Finally, local, grassroots opposition to nuclear energy and radioactive waste is found in (potential) disposal communities, where a broad range of local stakeholders work to keep their communities and environments from being turned into the nation's radioactive waste dump.

For example, in 1989 the Nuclear Regulatory Commission (NRC) began review of a proposal submitted by Louisiana Energy Services (LES) to build the nation's first privately owned uranium enrichment plant in Claiborne Parish, LA. The proposed site was located one-fourth of a mile from the town of Center Springs and one and one-fourth miles from the town of Forest Grove. Both were predominately African American communities founded in the 1860s, and both had poverty rates above 60 percent (Bullard, 2000, pp. 130). Residents from these cities and the nearby city of Homer organized themselves into a group called Citizens Against Nuclear Trash (CANT) and challenged the site selection process and proposed outcome. They acquired the services of the Sierra Club Legal Defense Fund (which has since changed its name to the Earthjustice Legal Defense Fund) and sued LES. They claimed LES and the NRC staff were engaging in environmental racism. The lawsuit drug on for more than eight years, during which time CANT continued to argue that two African American communities had been rendered "invisible" since they were not even mentioned in the NRC's draft environmental impact statement (Bullard, 2000, pp. 131). In fact, only after sustained and intense public outcry did the NRC even attempt to address the obvious environmental justice and "disproportionate impact" implications. And even then the Final Environmental Impact Statement devoted less than one page to addressing the environmental justice implications of siting a uranium enrichment plant in a predominately poor, African American community.

Throughout the lawsuit, CANT presented evidence showing that as LES narrowed their site selection process, from national to the state of Louisiana to Claiborne Parish, the aggregate percentage of the black population within a one-mile radius of the proposed sites continued to rise—from 28 percent to 37 percent to 65 percent and finally 97 percent black population within a one-mile radius of the "LeSage" site on Parish Road 39 (Bullard, 2000, pp. 130–131). Overall, beyond claiming their communities had been rendered "invisible," CANT argued that the NRC's environmental impact statement was inadequate because it failed to accurately assess the costs and benefits of the proposed plant. More specifically, they argued the costs had not been analyzed in relation to the nearby communities where the disproportionate burden of health and safety, property values, accidents, traffic, noise, not to mention radioactive dust in the air and water, would all be most severe. Additionally, CANT argued both LES and the NRC had failed to consider (much less justify or legitimate) the inequitable distribution of costs and benefits to white and African American populations. Nor did either the NRC or LES acknowledge how the proposed action would follow a clear national pattern of "institutionally biased decision-making [that] leads to the siting of hazardous facilities in communities of color and results in the inequitable distribution of costs and benefits to those communities" (Quoted in Bullard, 2000, pp. 131–132).

After more than eight years, in May of 1997 the three-judge panel of the NRC Atomic Safety and Licensing Board issued a final decision finding "racial bias played a role in the selection process" and denied the permit. The NRC's decision was appealed by LES and upheld the following year. The claims made by CANT throughout this process clearly draw from and rearticulate the social justice paradigms of recognition, distribution, and representation. This is reflective of the actual recognitional, distributive, and representational injustices involved in this case. The misrecognition (or rather almost total lack of recognition) and misrepresentation of the people of Forest Grove and Center Springs within LES' proposal and the NRC' environmental impact statement facilitated the distributive injustice of siting a uranium enrichment plant in an impoverished community of color.

As has been argued throughout the present study, decisions about the production, distribution, and consumption of energy and electricity involve more than just technical/scientific considerations. This is because energy systems take on significant social dimensions as these systems are developed, contested, and implemented (Sze, 2005, 2007). This can be clearly seen by analyzing the "environmental justice energy activism" of antinuclear power organizations like CANT and others working to highlight these social dimensions of energy systems on the local level. As is detailed below, analyses of the claims made by local level antinuclear power organizations supplement the conclusions identified from the analyses of national level organizations. More precisely, the analysis of claims made by local level antinuclear organizations also reveals how these groups draw upon and reconstruct the social justice paradigms of distribution, recognition, and representation in their arguments against the financing of new reactor construction through "construction work in progress" (CWIP) charges to utility ratepayers, and in their struggles for the increased safety and security of nuclear facilities.

This chapter will show how the distributive, representational, and recognitional injustices associated with CWIP financing parallel those associated with the federal "loan guarantee" program on the national level. Both are mechanisms for socializing the financial costs/risks of constructing new nuclear reactors, while enabling investor-owned utilities to privatize any eventual financial benefits/profits. Analyzing the claims made by local level antinuclear organizations regarding this issue enables the identification of additional principles of energy justice that emerge from the local level standpoint. Following this, the analysis turns to how representational, recognitional, and distributive injustices are involved in local antinuclear organization's efforts to increase the safety and security of nuclear facilities. These efforts include campaigns to close unsafe reactors, prevent existing reactors from being relicensed and/or new reactors from receiving a license, and generally increasing the safety mechanisms and emergency preparedness for existing nuclear facilities. And once more, the analysis of how local level organizations draw from and reconstruct different discourses of justice not only helps highlight the nature of the multidimensional injustices associated with nuclear energy but also enables the identification of additional principles of energy justice.

CWIP FINANCING FOR NEW NUCLEAR REACTOR CONSTRUCTION

The claims made by local level antinuclear organizations in their arguments against CWIP charges draw from and reconstruct the social justice paradigms of recognition, representation, and most prominently the paradigm of distribution. Basically, the arguments made by local level antinuclear organizations against CWIP charges mirror those made by national level organizations with regard to the federal "loan guarantee" program. Given the complete unwillingness of private investors to invest their money in the construction of nuclear reactors, which are extremely long-term, high-capital, and high-risk projects, and given the fact that the $10 billion average price of a new reactor is beyond what any electric utility can afford to finance itself, the only way new reactors can be constructed in the United States is to publicly finance them with either taxpayer money or, in the case of CWIP charges, ratepayer money.

CWIP charges let utilities charge ratepayers for the cost of financing new power plants during construction, before they are completed and providing electricity. In the absence of CWIP financing, electric utilities are only allowed to charge customers for the costs of plants that are "used and useful" (i.e., actually in service and producing electricity reasonably required to meet the needs of customers). CWIP charges are often sought when utilities want to build power plants that Wall Street is reluctant to finance. Utilities understand that in a competitive marketplace, businesses have to keep their costs under control, and passing along unnecessary costs to customers is bad business. However, in the regulated monopoly settings that utilities generally operate, utilities are already insulated from that kind of marketplace discipline. CWIP charges thus let utilities avoid the process of using stockholder dollars and taking out loans to build power plants, and also lets them avoid the wait to collect their investments after plant construction. For these reasons, utilities favor CWIP charges, but its effect on consumers is much less positive. CWIP charges essentially convert consumers into captive and involuntary investors, placing the burden of up-front financing costs onto them. The costs end up on their bills sooner, before they ever receive electricity from the plant in question, and there is little recourse should the costs skyrocket or the project be abandoned. CWIP charges establish a perverse incentive structure for utilities by eliminating the incentives to keep costs low or pursue other means of fulfilling their mission besides constructing expensive power plants. In essence, they allow utility companies to gamble with their customer's money; if the project goes way over budget or is cancelled, the utility does not lose anything, if the project succeeds the utility retains the benefits. It is a real, "heads I win, tails you lose" situation between electric utilities and their customers.

Within the strategically significant period of time under consideration for this study, several local level antinuclear organizations in different states were engaged in campaigns against the establishment of CWIP charges to finance new nuclear reactors. One such organization was Missourians for Safe Energy (MSE). While there were other local antinuclear organizations working on this issue during this time, the arguments made by MSE, and discourses of justice that informed them, are representative of the arguments made by other groups in other locations (such as Iowa and Florida) around the country. MSE is an antinuclear and pro-safe energy organization based in Columbia, MO, with members from around the state. They are a 501(c) 3, nonprofit organization that engages in public education and advocacy for a carbon-free and nuclear-free energy future. Concerned citizens in Missouri organized themselves into MSE in the early 1970s in response to their anger over the construction of nuclear reactors in central Missouri. In 1976, Union Electric (the local utility that has since changed its name to AmerenUE) was using CWIP charges to finance the construction of two huge 1,150 megawatt reactors in Callaway County. Because the Public Service Commission (PSC) (a regulatory commission used by many states to oversee electric utility companies that are "natural monopolies") had approved CWIP charges both reactors were being constructed simultaneously. In that same year Missouri voters passed Proposition One by a 2-1 margin, which made CWIP charges illegal in the state. When the new law pulled the plug on ratepayer financing, UE first announced a delayed construction schedule for Callaway 2, and then several years later cancelled the project altogether. When Callaway 1 was completed in 1984 it wound up costing more than $3 billion, and UE went to the PSC asking for a 70 percent rate hike to pay for it. According to MSE, if CWIP funding had remained in place, it is likely UE would either have sunk hundreds of millions more into Callaway 2 before realizing that it was not needed or they would have actually completed it. Regardless, it was in fact not needed and consumers would have had to pick up the cost—hundreds of millions, if not billions—to pay for this unnecessary plant.

From the standpoint of local level antinuclear organizations, such as MSE, there are several interrelated injustices associated with ratepayer financing of new reactor construction. First, CWIP financing encourages overbuilding and removes important opportunities for democratic oversight. When CWIP financing is in place utilities loose the incentive to build only the capacity they expect to need, because they are no longer financially responsible for the risks/costs. Also, without CWIP charges, before the costs of a new plant are added to the utility's rate base and included in customers' bills, careful auditing of all expenditures is done. In the absence of CWIP charges there is the opportunity for the PSC staff, the Public Counsel (representing public interests before the PSC), and consumer groups to challenge the legitimacy of expenditures. But, with the introduction of CWIP charges, the opportunities for these types of prior oversight are removed. Thus, a representational injustice that limits democratic decision-making on the local level follows from the introduction of CWIP financing, which sets the stage for misrecognition of public interests in relation to the interests of investor-owned utilities. It follows from this analysis that energy justice requires mechanisms for facilitating public involvement at all relevant levels of decision-making to the greatest extent possible. Because decisions about energy systems in the United States are made on multiple levels (local, state, regional, national) and because the impacts of those decisions are felt on multiple levels, democratic decision-making based upon public input and participation is required at each level. This will help ensure the public interest on any one level is not sacrificed for those on another (or for private interest of capital accumulation).

Second, local level antinuclear organizations claim CWIP charges are patently unfair, and not only because they transfer the risk from investors to customers. Utility stockholders earn a rate of return on their investment commensurate with risk taking, but with CWIP charges the utilities' customers are actually be the ones assuming the risk by paying for plants that may never be completed or produce power. Nuclear utilities, such as AmerenUE, generally try to argue that CWIP charges are a "pay-as-you-go" plan that reduces future "rate shock." This argument suggests that it will *benefit* customers to pay for a new plant in advance, as it will reduce accrued finance charges, and thus the ultimate cost of the plant. However, as MSE and other local level antinuclear organizations have shown, this reasoning is severely flawed from a social and economic justice point of view. For one thing, paying in advance is inequitable to people who are paying now but will never receive power from the plant. This is especially relevant to the elderly and many other people, given the highly mobile nature of contemporary society that results in people frequently moving from one location to another. For another thing, CWIP charges also mean that customers who pay now are paying more than the nominal value of their money. Given the "time value of money," the present value of a dollar this year is greater than the value of a dollar ten years from now. If the extra money that utilities want ratepayers to pay now were to remain in their hands, it could be earning interest in a bank account or it could be invested in other financial assets. Alternatively, this money could be invested in making the ratepayer's homes or businesses more energy efficient (more insulation, better lighting, more efficient appliances, etc.). This would not only lower present utility bills, but it would also reduce future demand for additional power. In any case, local level antinuclear power organizations argue that ratepayers would be better off holding onto their money.

These general arguments made by local antinuclear organizations against CWIP financing for new reactor construction were employed and further specified by MSE in 2008, 2010, and 2011 when AmerenUE sought to pass legislation that would overturn Proposition One and allow the utility to use CWIP charges to finance an "early site permit" for a second reactor in Callaway County. In a letter from MSE that was delivered to Missouri State Senators dated January 31, 2011, MSE points out that while the legislation in question would only allow for CWIP financing to be used by AmerenUE to obtain an "early site permit," this would set an unfortunate precedent and essentially be the "proverbial 'nose of the camel under the tent.' If Ameren and their partners decide to attempt construction of a new reactor at Callaway, they would certainly be back asking [the state legislature] to authorize CWIP charges for all construction expenses" (MSE, Letter to Missouri State Senators, 1/31/2011). The cost of obtaining an "early site permit" is miniscule ($40–$45 million according to AmerenUE's estimates) compared to the massive costs of building a plant. The letter further argues, if the investment community deems this minimal expenditure imprudent, it is extremely unlikely that they would be willing to underwrite a project that may take a decade to complete and cost $10 billion or more.

Nuclear projects have historically been extremely risky, with a high cancellation and default rates, and huge cost overruns on plants that have been completed. Callaway 1, for example, was completed at more than six times its original cost estimate, and upon completion, Union Electric went directly to the PSC asking for a 70 percent rate hike. Even after adjusting for inflation, the completed reactors of the last generation ended up costing more than three times their initial cost estimates (MSE, Letter to Missouri State Senators, 1/31/2011). Nuclear power has a fifty-plus-year track record, and despite massive subsidies ($154 billion in 2007 from 1950 to 1990), it has repeatedly failed the test of the marketplace (IEER, 2007). This is what led *Forbes Magazine* in 1985 to describe the failures of America's nuclear power industry as "the largest managerial disaster in business history, a disaster on a monumental scale" (Cook, 1985). The MSE letter goes on to conclude, "the private sector rightly deems new nuclear plants too risky to finance. As such, utilities interested in new nuclear units demand the risk be transferred to the ratepayers or the taxpayers. This is patently unfair to Missouri families and businesses" (MSE, Letter to Missouri State Senators, 1/31/2011). This is especially true, according to MSE, given that investments in energy efficiency can eliminate the need for new plants at a tiny fraction of the cost, and considering that wind power is already significantly cheaper than coal or nuclear. These clean alternatives (energy efficiency and wind power) can be installed far more quickly—providing jobs and energy in the near term—and are certain investments, unlike cancellation and default-plagued nuclear projects (IEER, 2007).

As the conflict and the political process played out over the next several months, MSE received support and expressions of solidarity from national level antinuclear organizations such as NIRS. This provides a clear example of movement coordination and resource sharing between the different levels of the antinuclear movement in the United States. In fact, on February 16, 2011, NIRS sent a letter to their members and supporters on behalf of MSE and others in Missouri fighting the introduction of CWIP charges. The letter reflected and reiterated the arguments made by MSE, and similarly drew upon ideas of distributive justice to structure their positions. More specifically NIRS informed their members:

> The Missouri legislature is currently considering legislation that would force ratepayers to foot the bill for some costs related to the Callaway-2 nuclear reactor before it is built and operating. This violates the will of Missouri voters, who passed a ban on Construction Work In Progress (CWIP) charges by a nearly two-to-one margin. And while this legislation is limited in scope, it's just the foot in the door. If Ameren receives a permit to build this reactor, you can bet they'll be back looking in ratepayers' wallets for the $10 Billion or more it would cost to actually build the thing. (NIRS, Action Alert, 2/16/2011)

Active members of MSE were very appreciative of the support and involvement of NIRS in this local campaign.

It is important to point out that the expressions of solidarity and cooperation went both ways, between national and local level antinuclear organizations, with regard to these issues of public financing of new reactor construction. The parallels between the two approaches ("loan guarantees" and CWIP charges) and the multidimensional injustices associated with both were patently obvious to organizations working on both levels of the movement. Consequently, throughout the eighteen-month timeframe under consideration here, MSE worked to distribute and spread awareness of each and every action alert and petition NIRS and other national level groups put together regarding the federal loan guarantee program (discussed in previous chapters). "People power" is without question the most important resource the antinuclear power movement has at its disposal. Antinuclear organizations on all levels are fully aware that they can never hope to outspend the nuclear industry in terms of lobbying, public relations, and so on. Their best hope is to mobilize and educate as many people as possible to make their voices heard in the decision-making and political processes, in hopes that democratic principles will win out in the end. NIRS sending an action alert to their 20,000+ supporters informing them of what MSE was fighting against and asking them to make their voices heard was a significant contribution of social/network resources that helped ensure the legislation eventually stalled.

While the movement's multiple framings of the issues allowed national level groups to form a coalition with other fiscally conservative groups like American for Generational Equity (discussed in previous chapters) against the "loan guarantee" program, MSE was able to similarly form a coalition with other environmental organizations who generally oppose nuclear power (such as Missouri Coalition for the Environment) and other large utility customers who generally did not take a position on nuclear power. MSE's most significant ally was Noranda Aluminum Holding Corporation which operates a large aluminum smelting plant in Southeastern Missouri. Noranda is the largest industrial consumer of energy in the state, and they were motivated to get involved with the CWIP issue out of fear that the additional energy costs would hurt their profitability. Noranda was able to devote significant financial resources to the conflict, helping to pay for state-wide television advertisements encouraging Missourians to contact their legislators to tell them to vote against CWIP financing. Additionally, because Noranda is the largest industrial electricity consumer in the state who employs large number of people, they were able to exercise leverage in negotiations with AmerenUE and were able to put meaningful pressure on state legislators in Jefferson City. In sum, the multidimensional injustices associated with federal "loan guarantee" program enabled national level antinuclear organizations to frame the issue in different ways, drawing from interrelated paradigms of justice, which allowed for a coalitions to form between antinuclear groups and other organizations not generally concerned with nuclear power. In much the same way, MSE and other local level antinuclear organizations were able to frame the problems of CWIP financing in different ways, again drawing from the interrelated paradigms of social justice, which enabled them to form a coalition that included large industrial users of electricity, which were generally not concerned with nuclear power per se, but were very concerned about the distributive injustices CWIP financing would produce.

These analyses have shown that while issues of distributive (in)justice were paramount in both cases, problems of representational and recognitional injustices also emerge, reflecting additional dimensions of the injustices involved. These additional dimensions of the injustices associated with the public financing of new reactor construction, specifically through CWIP charges, became especially apparent following the multiple reactor meltdowns at Fukushima in March 2011. For instance, two weeks following the onset of the Fukushima disaster MSE delivered another letter to the Missouri State Legislature that reiterated their earlier arguments against the partial repeal of the law banning CWIP charges and added the claim, "Fukushima Daiichi means absolutely no private investment in new nuclear" (MSE, Letter to State Legislatures, 3/25/2011). MSE again pointed out the fact that after losing tens of billions in the "great nuclear crash" of the 1980s, Wall Street has refused for nearly three decades to consider financially risky new nuclear plants as an acceptable investment option. And while the industry had hoped that a so-called nuclear renaissance would change this, dramatic cost escalations and utilities backing away from or putting off new nuclear projects has remained the norm. And most significantly, "any remaining hopes for a market-based turnaround went out the window with the multiple meltdowns in Japan" (MSE, Letter to State Legislature, 3/25/2011).

MSE predicted (and were latter proven correct) that following the disaster in Japan,

> We will surely hear from the industry that new nuclear units are safer. The reality, however, is that nuclear power presents unique catastrophic risks. While the likelihood of occurrence can be reduced, there are still a large variety of disaster scenarios. And given that each reactor core contains the radiological equivalent of 1,000 Hiroshima bombs, the potential losses are astronomical. (MSE, Letter to State Legislature, 3/25/2011)

Thus nuclear disasters represent a case of what Freudenburg and Gramling (2012) refer to as "low-probability high-consequence" events that existing regulatory mechanisms are notoriously awful at recognizing and addressing. Former NRC commissioner Peter Bradford highlighted the connections between the "unique catastrophic risks" of nuclear power and investor's unwillingness to get involved in nuclear projects. Speaking with regard to the Three Mile Island accident of 1979, Bradford said, "Wall Street learned that a group of licensed operators no worse than any other could transform a billion dollar asset into a two billion dollar clean-up in ninety minutes" (Quoted in MSE, Letter to State Legislature, 3/25/2011). The costs of the Fukushima disaster will dwarf those associated with Three Mile Island. The economic losses from the meltdowns at Fukushima are estimated to be in the tens of billions and could easily reach into the hundreds of billions by the time some semblance of a solution is in place. This means because of the "unique catastrophic risks" associated with nuclear power, it is virtually certain that private investment capital will not back any new nuclear projects now or in the foreseeable future. In other words, because the production of nuclear power has the potential to create such dramatic environmental as well as social, economic, and political damages, the market simply will not get involved in nuclear development. It seems almost ironic that not only are ratepayers and/or taxpayers being unjustly forced to pay for energy development project that are financially hazardous, but these same projects also carry the potential to do almost incalculable environmental and social, political, and economic damages that would negatively impact these same ratepayers and/or taxpayers.

Local level antinuclear organization's arguments against CWIP financing clearly incorporate ideas of distributive, recognitional, and representational (in)justice. Beyond the fairly obvious distributive injustice of saddling ratepayers with the risks/burdens of financing new reactors, while investors retain the eventual profits, other less obvious distributive injustice also emerges. CWIP financing of new reactors is inequitable not only for those who are paying up-front for a product/service they may never receive, but also for anyone who could use the same money for any number of more useful/beneficial purposes. Beyond these obvious and less obvious distributive injustices, the analysis of antinuclear groups' arguments reveals the roles of representational and recognitional injustices as well. The establishment of CWIP charges necessarily takes away important avenues for public oversight and protection with regard to the development and implementation of energy systems, by essentially granting utilities a blank check to proceed with development projects. This all but removes the role of public input into decisions about energy system development, which has been shown to lead to the misrecognition of public/consumer interests at the expense of capital accumulation. This is all compounded by the fact that the distributive injustices at work here are being used to facilitate the development of an energy system (nuclear power) that carries "unique catastrophic risks," which could potentially harm the same people being cajoled into financing their development.

Therefore, following from this analysis, it becomes clear that energy justice requires decisions about the public financing of energy systems, which must include mechanisms that protect the interests of the public both as citizens/taxpayers and as consumers/ratepayers. As should be clear, the antinuclear movement's claims against federal "loan guarantees" on the national level and CWIP financing on the state/local level share many important similarities with regard to the distributive, representational, and recognitional injustices involved. This is unsurprising considering both "loan guarantees" and CWIP charges are essential mechanisms for socializing the risks of energy system development. What is different is the subject position of who is victimized by the distributive, representational, and recognitional injustices. Therefore, the conclusions derived from the analysis of local level opposition to public financing of new reactors add additional considerations and dimensions to the principle of justice derived from the analysis on the national level. This supports Fraser's (2009) contention that the "scale" of justice is important when trying to theoretically or empirically determine what constitutes just social arrangements. In other words, because conflicts over social, environmental, and energy justice in the United States take place on multiple scales (local, regional, national), theories of justice in any of these domains need to be sensitive to the injustices that can arise on or between these levels.

LOCAL EFFORTS TO INCREASE THE SAFETY OF NUCLEAR REACTORS

Campaigns to increase the safety of nuclear facilities, including campaigns to close unsafe and unneeded nuclear reactors, flourished in the United States on both the national and local levels and included significant organized protest in the 1970s. Perhaps the most famous of these protests were the Clamshell Alliance's protests at the Seabrook Station Nuclear Power Plant in New Hampshire and the Abalone Alliance's protests against the Diablo Canyon Nuclear Power Plant in Southern California. Both the Abalone Alliance and the Clamshell Alliance are local antinuclear organizations formed in the 1970s to oppose the construction of nuclear reactors in their states. Thousands of antinuclear demonstrators were arrested at both locations in the late 1970s while engaging in nonviolent, direct actions, including some 2,000 members of the Clamshell Alliance occupying the Seabrook Nuclear Power Plant in 1977. In addition to these notable local level protests against nuclear power, there were also significant national level protests against nuclear power, especially following the Three Mile Island (TMI) accident in 1979. More specifically, 65,000 people including the Governor of California demonstrated against nuclear power in Central Park in New York City in May of 1979. And in September of that year, almost 200,000 people attended a protest against nuclear power in Washington, DC (Giugni, 2004). These local and national protests over the production of nuclear power and radioactive waste received significant media and public attention.

However, within just a few years, by the early 1980s much of the antinuclear activism in the United States pivoted from a focus on nuclear power and radioactive waste to the issues of nuclear weapons. For instance, On June 12, 1982, one million people demonstrated in New York City's Central Park against nuclear weapons and for an end to the cold war arms race. It was the largest antinuclear protest and the largest political demonstration in American history (Schell, 2007). There are at least two reasons why the antinuclear movement in the United States switched focuses so rapidly in the early 1980. First, the unsustainable economic model of the nuclear industry was becoming readily apparent by this time and had been utilized by the antinuclear movement throughout the 1970s. As discussed above, major cost overruns and construction delays were endemic within the nuclear industry "boom years" of the 1960s and 1970s (Cook, 1985). In fact, much of the antinuclear movement's political strategy during the 1970s was to work to delay and then further delay the construction of nuclear facilities, knowing the economics become more unfavorable the longer they could stall the project. So, by the early 1980s, with the assistance of the antinuclear movement, the nuclear industry was collapsing under the weight of their own incompetence. The other major reason for the switch in focus had to do with the election of Ronald Reagan in 1980, who immediately after taking office in 1981 began escalating the U.S. nuclear weapons program to new heights not seen before or since. This along with the corresponding spike in Cold War hostilities, the strength of the broader peace movement of that era in the United States drew many antinuclear power activists and organizations into the pressing conflict over nuclear weapons reduction (Joppke, 1991).

Having said all this, campaigns to increase nuclear safety and to close nuclear reactors in the United States continued through the 1980s, 1990s, and the first decades of the twenty-first century. These included successful campaigns, like the one that eventually succeeded in closing the Yankee Rowe Nuclear Power Plant in Vermont, as well as the Crystal River Nuclear Power Facility in Southern Florida. The event that provided much-needed fuel for these campaigns to increase reactor safety in the United States, as well as a renewed public and media visibility, was the multiple reactor meltdowns at Fukushima Diachii in Japan on March 11, 2011. In the weeks and months that followed the onset of this ongoing tragedy, existing campaigns to increase reactor safety in the United States took on new urgency. Local antinuclear organizations and activists found themselves with an audience, and a level of general public interest, that they had not experienced in decades. Furthermore, new campaigns for increased nuclear safety and new antinuclear organizations were emerging around nuclear facilities (especially nuclear reactor facilities) across the country.

Analyzing the claims for increased nuclear safety made by local antinuclear organizations in the wake of the onset of the Fukushima disaster, including claims for the closing of unsafe reactors, shows how ideals of representational, recognitional, and distributive (in)justice underlie and are reconstructed through social and political conflicts over energy production. More specifically, the operation of unsafe nuclear facilities is based upon representational injustices within the operations of the NRC and state-level regulators that prevent sufficient public participation in decisions about the (continued) operations of these nuclear facilities. The analysis of the claims made by local antinuclear organizations shows how these representational injustices lead to the unjust misrecognition of public concerns and interests, which ultimately facilitates the maldistribution of benefits and burdens between public interests (local communities, citizens/taxpayers, consumers/ratepayers) and private interests of capital accumulation (the nuclear industry including investor-owned utilities). And most significantly, identifying these empirical and normative injustices establishes possible principles of a framework for energy justice. Taken together, the framework these principles form marks an important step toward establishing a sociologically informed basis for evaluating the (un)desirability of different energy policies and energy systems.

One such campaign that was already well underway by the onset of the Fukushima meltdowns, but gained considerable attention and momentum following the disaster, was the campaign to close the Vermont Yankee Nuclear Power Plant located on the Connecticut River outside of Vernon. While there were other similar campaigns to increase the safety of nuclear facilities and close unsafe reactors around the United States, the case of Vermont Yankee (VY) is representative of these kinds of conflicts/campaigns in some respects, while being unique in regard to the large amount of antinuclear organizational activity and eventual outcome. More specifically, it is typical of campaigns to close unsafe nuclear facilities in that the claims made by local level antinuclear organizations express and reflect the importance of representative, recognitional, and distributive justice. This highlights the central problems with misrepresentation of public concerns/desires within the NRC leading to the misrecognition of these and ultimately the maldistribution of both social and environmental benefits and burdens.

One such local antinuclear organization that was very active in the conflict over VY was Vermont Citizens Action Network (VTCAN). VTCAN is a grassroots antinuclear, pro-safe energy organization pushing to maximize democratic decision-making and control over the decision to close VY and establish a clean, renewable energy system. In a factsheet about where the waste from VY goes that was produced by VTCAN during the conflict over the closure of VY, many of the interrelated injustices associated with the operation of this reactor (and all other reactors) are highlighted, including the environmental injustices in different levels and in different locations. To answer the question, "where does the waste from Vermont Yankee go?" VTCAN first points out "The waste stream is long and wide. While much of Vermont Yankee's environmental impact is in the tri-state area (Vermont, Massachusetts and New Hampshire), it starts and ends far from our small corner of New England" (VTCAN, Nuclear Waste Factsheet).

The organization goes on to highlight how as part of its routine operation, Vermont Yankee vents radioactive material into the air through its cooling system on a routine basis. Many of the gases released break down into radioactive particles that settle in surrounding communities as a kind of "fall-out." According to VTCAN, Vermont Yankee has released over 400,000 curies of radioactive waste into the air during its thirty-six-year history.

In addition to these "routine releases" of radioactivity into the air, VTCAN claims VY has released more than an additional 2,000 curies of radioactivity into the Connecticut River, in the form of tritium and other radioactive contaminants released by the plant's cooling system. Furthermore, these environmental degradations correspond with environmental injustices accumulating locally, especially in the town of Vernon, VT. Due to Vernon's proximity to VY, the residents of this town suffer these negative consequences to a greater degree than do other populations, as well as the increased risks associated with the over 5,000 tons of high-level radioactive waste stored seven stories in the air at the site (VTCAN, Nuclear Waste Factsheet). However, what is most significant about VTCAN's claims and arguments about the interrelated injustices of nuclear power production in New England is their explicit acknowledgment of the *other environmental injustices* that victimize other people in other places as a result of the operations of VY. This clearly shows how issues of (in)justice are of central importance with regard to the conflicts over nuclear power and other dirty, dangerous, and expensive energy systems. VTCAN points out that:

> "Low-level" radioactive waste, which includes all waste except the used fuel, is shipped to Barnwell, SC. Barnwell is a poor, rural, 48% African American community that hosts the country's primary radioactive waste dump. Although the dump is slated to close in 2009, within a few years the community's water supply will become contaminated by waste leaking from the dump, raising concerns about environmental racism. After the Barnwell dump closes, more radioactive waste may have to be stored in Vermont unless another community is forced to host a nuclear waste dump. (VTCAN, Nuclear Waste Factsheet)

This clearly reflects the organization's awareness of the fact that the operations of this and all other nuclear power facilities do not just create distributive, recognitional, and representational environmental injustices for local stakeholders, but also for other poor and/or otherwise marginalized communities in many different locations. This cascading flow of environmental and energy injustices is by no means limited to the shipment of so-called low-level radioactive waste to South Carolina. For instance, VTCAN goes on to note how:

> Contaminated uniforms, gloves, and booties are sent off-site to be "cleaned" at industrial laundries that serve the nuclear industry. The closest such laundry is operated by UniFirst, Inc. in East Springfeld, MA, a largely minority and immigrant community. These laundries routinely have bad safety and working conditions, fail to train their employees about radiation hazards and discharge radioactive and chemical waste into the local water supply. The UniFirst/NTS laundry in East Springfeld has repeatedly dumped waste in the local sewage system and even the pond in a nearby park. Workers have been endangered by fires and spills in the plant. (VTCAN, Nuclear Waste Factsheet)

Furthermore, and just as significantly for the present study, VTCAN links the national and local levels of antinuclear power movement in the United States through issues and a framework of environmental (in)justice when they argue,

> The beginning of the "nuclear fuel chain" is the mining and refining (called milling) of uranium ore. Mining and milling operations produce immense amounts of radioactive and chemical waste. They are mostly located on Native lands in the Dakotas, the Province of Ontario and the Southwest. For every pound of uranium that is used in a reactor, 3,500–4,000 pounds of radioactive uranium tailings are generated. (VTCAN, Nuclear Waste Factsheet)

These and other claims made by VTCAN (and many other similar claims made by other local and national antinuclear organizations) clearly show that ideas of justice, fairness, rights, and entitlements are powerful motivators that propel these organizations in their fight against unjust power systems. This provides important validation of the theoretical and conceptual approaches utilized in the present study. Given the importance of ideas about justice demonstrated above, it is important to examine how these organizations draw from and reconstruct the multiple and interrelated dimensions of justice associated with ideas of distribution, recognition, and representation.

The stage was set for the conflict over VY back in 2002 when Louisiana-based Entergy Corporation purchased the plant. In exchange for state approval of the purchase, Entergy signed a memorandum of understanding agreeing that in order to renew its operating license beyond its original expiration in March 2012, they would need legislative approval from the state (Blanding, 2012). However, beginning in 2007 a series of troubling events began to erode public confidence in the facility and in Entergy's ability to operate it safely. In 2007, one of the plant's hourglass-shaped cooling towers suddenly and unexpectedly collapsed due to structural deficiencies in its construction, throwing water and debris into the Connecticut River. Following this, in the same year, Entergy attempted to spin off VY and five other plants into a highly leveraged company to be called Enexus. This move was decried by both the public and Vermont lawmakers for fear the new highly leveraged company would not be in a position to operate the plant safely nor fully decommission the plant once it was eventually closed (Blanding, 2012). The final straw that all but totally removed whatever public support remained for VY came in January 2010 when Entergy announced they had discovered radioactive tritium leaking into the groundwater surrounding the plant. Tritium is a radioactive isotope of hydrogen, which mimics its nonradioactive counterpart, and thus is easily absorbed into the human body. Entergy officials first testified under oath that there were no underground pipes below the facility from which the waste could leak. Shortly afterward it was found, and the company eventually admitted, that such pipes did in fact exist, and have been leaking tritium since at least 2005 (Blanding, 2005). Tests later showed the levels of contamination were below those considered harmful to humans, according to the EPA's and the NRC's guidelines, but the damage to Entergy's dwindling credibility was already done.

This episode provides clear evidence suggesting the foundational problem in this case, and all other conflicts over increasing the safety of nuclear facilities has to do with the lack of public confidence in the NRC's willingness and/or ability to effectively regulate the production of nuclear power in the interests of protecting public health and the environment. After all, local antinuclear organizations, other citizen groups, and eventually the state legislature (discussed below) would not have to fight to close VY if the NRC was adequately monitoring the plant and preventing events like cooling towers collapsing and radioactive water leaking from submerged pipes. Of course this last statement assumes nuclear power plants can, in fact, be adequately monitored so as to actually protect public health and the environment, which is by no means certain.

Recent independent monitors and other watchdogs, including both local and national antinuclear organizations, have shown the NRC has moved away from enforcement of regulations and has "fallen down on the job." Despite having a staff of over 4,000 the NRC audits only 5 percent of activities at plants in the United States in a given year (Blanding, 2012). Since the year 2000, the NRC has approved sixty-two out of sixty-two applications it has received to relicense existing reactors for an additional twenty years beyond their initial thirty-forty year projected lifespan. This is despite mounting evidence that especially "first-generation" plants built in the 1960s are being dangerously approved for operation beyond what their design and constitutive materials can handle (IEER, 2007). In fact, according to a report by the NRC's own inspector general, the NRC had done little more than rubber-stamp the process by literally cutting and pasting, word for word, information from nuclear companies' applications into its approvals (Blanding, 2012). In June of 2011, the *New York Times* reported the number of civil penalties issued by the NRC to nuclear operators had dropped by 80 percent since the late 1990s. Even worse is the fact that the same month the AP released a report that found when plants have failed to meet safety and performance guidelines, the standard response from the NRC has been to lower its standards, through "pencil engineering," to get the plants back into compliance, rather than issuing penalties and forcing nuclear operators to abide by guidelines put in place to protect public safety and the environment (AP, "Safety Rules Loosened for Aging Nuclear Reactors, 6/11/2011"). According to the Union of Concerned Scientists (UCS), nearly half of all reactors fail to meet fire safety guidelines enacted in 1980, and only 60 percent are in compliance with "voluntary" standards on groundwater pollution. Given this, it is not surprising that the AP reports 75 percent of nuclear reactors in the United States have leaked radioactive tritium into groundwater (AP, "Safety Rules Loosened for Aging Nuclear Reactors, 6/11/2011").

Returning to the case of Vermont Yankee, responding to overwhelming and sustained public outcry, the State Senate denied approval for VY's license renewal in February 2010. Entergy immediately sued, claiming the state lacked the authority to close the plant since it was doing so on the grounds that the plant was unsafe. Entergy's lawyers argued that according to federal law, only the NRC has the authority to close a nuclear power plant because of safety concerns (Blanding, 2012). During the trail Entergy's lawyers showed clips of Vermont lawmakers discussing and debating the issue, in which they "bent-over-backwards" to avoid using the word "safety" to avoid being preempted by federal law. Instead they referenced "reliability" concerns, "economic effects," and "environmental concerns," which Entergy's lawyers argued was nothing more than a pretext for the main concern of safety. One of the features that makes this case unusual is that the state legislature in this case was supportive of local antinuclear organization's concerns about the safety of the plant. In many other cases (New York, California, Florida, Michigan) state legislatures at best are divided on the issue of nuclear power, and at worst fully support the development of nuclear facilities in their state in accordance with "growth machine" politics that see any development projects (especially really *big* development projects like nuclear facilities) as inherently good, despite any negative social or environmental consequences. It is ironic that in this case state legislators actually concerned with protecting the health and safety of the people and environment were forced to employ other, auxiliary framings of the problems with nuclear power production. Again, the multiple and interrelated problems with nuclear power production, and the multiple and interrelated injustices that they embody, enable antinuclear activists and *politicians* (in this case) to frame the issues in strategically different ways, which proves to be an important resource.

One of the primary reasons state legislators in Vermont took such a strong and unusual stand against the nuclear power industry is local level antinuclear organizations and grassroots activists were relentless in their efforts to make their voices heard. This included multiple nonviolent, direct action campaigns aimed at Entergy, which included hundreds of protesters being arrested at the VY facility, as well as at Entergy's headquarters in New Orleans, LA. For example, on January 18, 2011, fourteen women of the Shut It Down Affinity Group were arrested for blocking the driveway at VY while participating in walking meditation to morn cancer deaths caused by radioactive emissions. The arrest marked the fourteenth time since 2005 that women from this group have been arrested for engaging in nonviolent, direct action at either the VY nuclear power plant or Entergy's local headquarters in Brattleboro, VT, to shut down Vermont Yankee (Shut It Down Affinity Group, Press Release, 1/20/2011). Prior to their arrests, the women participated in a walking meditation while remembering friends, family, and all who have died from or suffered from cancer caused by radioactive emissions. The participants wore black clothing and donned white death masks during the procession. Following their release, the women released a statement that read:

> No corporation has the right to poison our air, water, environment, and the future of humanity. We come to Vermont Yankee today representing the harbingers of death that emanates from this reactor. . . . We cry for our children, their children, and those many generations that will have to suffer from the poisons that are emitted from Vermont Yankee. . . . We call on the state to act responsibly and join us in shutting down Vermont Yankee NOW! We call on all persons who do not want to see their air, water, environment, and life killed by this nuclear reactor to join us in shutting it down by such citizens' non-violent actions now. Don't wait-tomorrow may be too late! (Shut It Down Affinity Group, "Press Release, 1/20/2011")

Local level antinuclear groups such as the Shut It Down Affinity Group, Vermont's Safe and Green Campaign, the Vermont Yankee Decommissioning Alliance, VT Citizens Awareness Network, and others organized meetings and strategy building sessions in order to facilitate a "People's Campaign to Close Vermont Yankee" (People's Campaign to Close Vermont Yankee, "Open Letter," 8/15/2011). In an open letter to local activists and concerned citizens, written by these organizations inviting people to attend a planning session in August 2011, the local antinuclear movement's position on closing Vermont Yankee was fully expressed. Claims for representative, recognitional, and distributive justice are woven together throughout the organizations' positions.

The letter begins by stating the purpose of the meeting is to

> develop a *people's campaign of nonviolent direct action* to shut down Vermont Yankee . . . Vermonters and our tri state community have worked for 20 years to shut down Vermont Yankee and are committed to a green energy future. Our voices were resoundingly heard when the Vermont Senate voted overwhelming to replace Vermont Yankee in 2012. Now Entergy is trying to undermine the will of the people and steal our vote away. (People's Campaign to Close Vermont Yankee, "Open Letter," 8/15/2011 emphasis in original)

They supported the state of Vermont in its efforts to fight Entergy, but they also felt it was important to let Entergy, the courts, the media, and the world know *the people* of Vermont and surrounding communities in New Hampshire and Massachusetts would not allow a Louisiana-based, profit-driven energy company to "subvert democracy and imperil the future of the New England region."

The connections between the subversion of democracy and representative injustices, on the one hand, and the misrecognition of people's interests and desires and the maldistribution of benefits and burdens (between private and public interest), on the other, is made explicate as these organizations layout the rational for their antinuclear activism:

> We, the People, cannot allow a mega-corporation like Entergy to subvert democracy and continue irradiating our region while adding to the tons of high level nuclear waste stored high above the banks of the Connecticut River. The people of this region—by means of petitions, letters, public hearings as well as demonstrations, and votes by the Vermont legislature and annual town meetings—have long and repeatedly expressed their will to close Vermont Yankee and replace its power with safe, renewable alternatives. Many nations—including Germany, Italy, Switzerland, and Japan—are taking action to halt the continued use of nuclear power. Vermont has done the same. (People's Campaign to Close Vermont Yankee, "Open Letter," 8/15/2011)

The post-Fukushima context emerges later in the letter from the organizations' argument:

> In the aftermath of Fukushima, there is no longer any doubt that nuclear power is an imminent threat to our lives, health, environment, and livelihoods. Vermont Yankee, approaching the end of its 40-year operating license, and running at 120% of its original design capacity, is an aging Mark-I reactor, identical

in design and age to the reactors that exploded and melted down in Fukushima. The inadequacy of the Mark-I design has long been known, yet the Nuclear Regulatory Commission has allowed 23 of these reactors to continue to operate in America. It is time to close these dangerous reactors, starting here and now with Vermont Yankee. (People's Campaign to Close Vermont Yankee, "Open Letter," 8/15/2011)

The claims made by these local antinuclear organizations clearly draw from, reconstruct, and recombine ideas reflective of the social justice paradigms of representation, recognition, and distribution. These claims emerge from over twenty years of real-world struggle against nuclear utilities and federal regulators to increase the safety of nuclear facilities, and to close this unsafe reactor. Analysis of these claims along with an analysis of the relevant contexts helps identify the real-world injustices that plague the operation of unsafe nuclear facilities, in this case and many others around the United States and the world. For instance, the NRC's readiness to grant VY a twenty-year license extension despite the repeated expressions of the interests of the people of Vermont—through petitions, letters, public hearings, demonstrations, and votes by the state legislature—clearly exemplifies a lack of democratic decision-making and thus significant representative injustice. But, as these local antinuclear groups' claims bring to light, democracy is not just being *passively* subverted here, it is being actively subverted by Entergy, a "mega-corporation" who along with the rest of the nuclear industry use their tremendous resources to gain undue influence over the political and regulatory process. Therefore, the concerns of the people in this case are institutionally misrecognized at the federal level by NRC, which enables the distributive injustice of radioactive waste to continue to accumulate on the banks of the Connecticut River. This is a distributive injustice in at least two respects. It is a distributive injustice in relation to the accumulation of radioactive waste at the site, which burdens local communities with the proximate risks of contamination, safety, environmental degradation, property devaluation, and so on. It is also a distributive injustice in terms of the interests of out-of-state corporate actors being advanced at the expense of the interests of local stakeholders.

Furthermore, as argued throughout this study, identifying the existence and extent of these multidimensional injustices allows for the proposal of features for a model of energy justice. Here, the analysis of the claims made by local antinuclear organizations in the campaign to close VY (above) suggests at least two additional principles of energy justice should be considered. The first follows from the analysis of the above conflict in the post-Fukushima context, and is in fact one of *the central* lessons that should be derived from Fukushima and other energy disasters (TMI, Chernobyl, BP's Deep-Water

Horizon blowout, etc.). First, energy justice requires that energy systems be developed in such a way as to reduce the possibility of catastrophic accident, even so-called low-probability, high-consequence catastrophes, to the greatest extent possible. Given that such events, like nuclear reactor meltdowns or deep-sea oil rig disasters, are by definition "low-probability" it is not surprising that regulators and industry have been shown to become complacent with regard to such scenarios. However, it has also been shown that often the probability of these "low-probability, high-consequence" events proves greater than industry or regulators initially believed to be the case. Additionally, it has often been the case that the consequences of these disasters are felt most severely by poor and otherwise marginalized communities, which represents multiple injustices in its own right. Safe and clean energy systems are increasingly becoming a possibility, and energy justice requires that they be developed to the greatest extent possible.

The second principle of energy justice that follows from the above analysis has to do with the foundational problem within the conflict to close VY, and all other such conflicts to increase the safety of nuclear reactors in the United States. Beyond the operational incompetence of Entergy Corporation to effectively and safely manage the facility lies the more central and general problem of the NRC unwillingness and/or inability to effectively and adequately regulate the nuclear industry in the service of protecting public interests, public health, and the environment. Given the multilevel and interrelated social, political, and economic dimensions of energy systems (that have been the basis of the present study) and the vital role that energy plays in contemporary society, energy justice requires that the implementation, production, and distribution of energy systems must include strong, effective, and independent regulatory oversight. More specifically, energy justice requires concrete mechanisms exist to ensure "regulatory capture" does not allow the industry to be regulated effectively taking control of the regulatory body itself. This could include procedures to stop the "revolving door" between industry insiders, lobbyists, and regulatory bodies, and/or potential third-party audit/review of the regulator's record with regard to public and private interests.

Both of these additional principles of energy justice that follow from the above analysis, along with the previous principles identified earlier in this study, will be further fleshed-out and expanded upon in the concluding chapter. By expanding and further specifying these features of energy justice proposed in this study, it is possible to construct a model for conceptualizing the relationships and the flow of different dimensions of injustice that are associated with the continued production of nuclear energy. While this model is by no means definitive, it does help illuminate the relationships between the multidimensional and interrelated injustices associated with nuclear and other dangerous forms of energy production. The goal in developing such a framework of energy justice is to move toward establishing sociologically informed arguments for/against the development of alternative energy system and/or energy policies. Clearly this framework was derived through an analysis of antinuclear organization's claims and arguments *against* nuclear power. Therefore, in order to establish the utility of this framework for future decision-making, it is necessary to apply these principles of energy justice to the case *for* expanding the use of renewable energy sources such as solar, wind, and geothermal.

Conclusion

As has been shown throughout the present study, decisions about the production, distribution, and consumption of energy and electricity involve more than just technical/scientific considerations. Energy systems take on significant social dimensions as these systems are developed, contested, and implemented (Sze, 2005, 2007). This can be clearly seen by analyzing the "environmental justice energy activism" of groups working to highlight these social dimensions of energy systems, such as antinuclear organizations. The analysis of national and local level antinuclear organizations' claims has focused on issues regarding the public financing of new nuclear construction (through federal "loan guarantees" or CWIP charges), as well as issues of the management of high-level radioactive waste and other campaigns to increase the safety of nuclear facilities. Throughout these analyses it has been shown how ideas of (un)just distribution, recognition, and representation help structure, and are reconstructed through, the arguments made by antinuclear groups against the production of nuclear power. Furthermore, these discursive investigations, combined with analyses of the historical and social, political, and economic contexts from which the movement's claims emerge, reveal the existence and extent of actual, multidimensional, and interconnected injustices associated with the production of nuclear power and other dirty, dangerous, and expensive energy systems. Most significantly, following the methodology utilized by Fraser (2000, 2003, 2009) by examining claims made in existing struggles against existing injustices as the starting point for the development of theories of justice (including energy justice as well as other types) we can begin to understand the full structural and normative extent of both just and unjust social arrangements. This has led to the identification of eight to nine principles of what together form a framework for energy justice. This framework could and should be used to guide future energy policy and decision-making. The goal of constructing this framework for energy justice is to take significant steps toward establishing sociologically informed arguments for and against the adoption of particular energy system and other types of energy policy. This will ultimately enable future choices about the production, distribution, and consumption of energy to be made in socially and environmentally beneficial ways.

The conclusion will, first, review the analyses that facilitated the identification of various features or principles of energy justice, as well as further specify and elaborate on the principles themselves. This will be done in the context of continued debate about the role of nuclear power in reducing greenhouse gas emissions and curbing climate change. Second, by expanding and further specifying the proposed features of energy justice it is possible to develop a model for conceptualizing the relationships and the flow of different dimensions of injustice that are associated with the continued production of nuclear energy and other unjust energy systems (figure 4.1). While this model is by no means definitive, it does help us think about the relationships between the multidimensional and interrelated injustices associated with nuclear power and other dirty, dangerous, and expensive forms of energy production. Together this provides practical and applicable suggestions as to how energy systems could and should be pursued and developed moving forward.

Third, questions of morality and ethics will be addressed, along with a discussion of the potential for future innovations in nuclear and other energy technologies. Questions of justice inevitably come into contact with questions of ethics—the ways people can tell right from wrong, or good from evil. The present study goes beyond simply identifying principles of energy justice by arguing these principles "should" be used to guide future choices about energy systems and technologies. Thus, a brief discussion of the ethical argument is provided showing that historical struggles also reveal ethical standards, as they do principles of justice. This feeds into a discussion of innovation in energy technologies and ideas for addressing the climate crisis and shows the principles of energy justice developed here have utility for questions of future investments in research, development, and deployment.

Finally, potential questions of "why energy justice" and "how are the principles of energy justice distinct from the Principles of Environmental Justice" are addressed. This includes responses to possible critiques of Fraser's framework and the associated liberal political categories of distribution, recognition, and representation that structure so much of the proceeding analyses. Additionally, potential limitations of the methodological approaches utilized here are acknowledged and addressed by highlighting the analytic advantages associated with the analysis of organizational claims-making during a limited timeframe. This discussion begins to point the way toward directions of future research that are needed to more fully develop and realize a theory of energy justice.

POTENTIAL PRINCIPLES OF ENERGY JUSTICE

As briefly mentioned earlier and presented in greater detail in the preceding chapters, the analyses of the U.S. antinuclear movement's claims made in their struggles against nuclear power show how these claims draw from and reconstruct ideas of distributive, recognitional, and representational (in)justice. The analyses of these claims, in combination with analyses of the historical and social, political, and economic contexts from which they emerged, allow for the identification of existing and interrelated distributive, recognitional, and representational injustices associated with the production of nuclear power and other dirty, dangerous, and expensive forms of energy production. Therefore, from a sociological and "energy justice" point of view, energy policy going forward should move away from nuclear power and toward more socially and environmentally just sources of energy and electricity. This is especially significant because the role of nuclear power in future energy choices to mitigate the effects of climate change is very much up for debate. Especially in the United States and Eastern Europe, some political-economic elites are misguidedly working to sustain and/or resuscitate nuclear power as a viable strategy in their efforts to increase "energy security" and reduce carbon emissions to combat the climate crisis. The principles of energy justice identified in this study demonstrate why all such efforts to force nuclear power into proposed solutions for climate change are mistaken, not only because of the associated injustices but also because these efforts necessarily take away resources (time and money) from other more socially and environmentally just solutions.

To begin with, the analysis of national-level antinuclear organizations' claims regarding the multidimensional injustices associated with federal "loan guarantee" program for financing new nuclear reactor construction, along with the historical and social, political, and economic contexts, leads to the identification of the first two (or three) principles of energy justice. The first principle of energy justice is: *energy justice requires that the development of energy systems should minimize up-front, capital-intensive requirements to the greatest extent possible*. In keeping with very basic market principles, energy systems should be pursued in accordance with the least-cost options that are available. This, of course, requires an honest calculation of the full costs of different energy systems, including the *externalized costs* associated with traditional coal, oil, and gas production, which levels the playing field for renewable energy sources like wind, solar, and geothermal. Additionally, these calculations should be done without consideration of the "sunk costs" associated with energy companies' existing *physical* infrastructure. The decisions regarding the development of this infrastructure were neither made democratically nor made to advance social and environmental well-being, which is the entire purpose of establishing principles of energy justice. This does NOT extend to the *social* infrastructure that has developed around fossil fuel and nuclear facilities. Workers and communities that are dependent upon nuclear and fossil fuel facilities must be fully represented in decision-making, have their needs recognized, and receive a just distribution of resources and opportunities. Also, given the positive correlation between time and money in the context of energy system development, it follows that *energy justice requires that decisions about the development of energy systems should also be made in accordance with the shortest possible timeframe for completion.* This is especially significant when decisions about energy systems are being made in the context of policy decisions to mitigate the climate crisis. Time is of the essence with regard to addressing the climate crisis. Virtually all serious climate scientists agree that humanity has either run out of time or is rapidly running out of time to head off the most disastrous consequences of climate change. Therefore, new energy strategies are needed that can be implemented as cheaply and quickly as possible.

The second (or third) principle of energy justice is: *energy justice requires that in cases where public financing of energy systems is necessary, because private financing is unavailable for whatever reason, the eventual profits/benefits should be retained publicly.* This could take the form of dividend payments to taxpayers/ratepayers or future credits toward their energy costs. It also follows from this analysis, and the analysis of local-level antinuclear organizations' arguments against CWIP charges, that a related (sub)principle of energy justice is: *energy justice requires that decisions about the public financing of energy systems must include mechanisms that protect the interests of the public both as citizens/taxpayers and also as consumers/ratepayers.* Ensuring that the potential profits/benefits are retained publicly by either taxpayers or ratepayers, as opposed to privately by investor-owned utilities, would significantly alter the incentive structure that leads privately held nuclear utilities to pursue such public financing in the first place. As is, nuclear utilities are encouraged to seek public financing for new reactor construction, because doing so shifts the risk of the up-front investment onto the public, while enabling the firm and its investors to control the potential profits, which are earned based upon the risk they are no longer taking. However, if the potential benefits were retained by the public, then the nuclear utility would lose the incentive to embark on the project in the first place, which could lead them to pursue more environmentally and socially, economically, and politically just energy systems that are cheaper and quicker to implement.

Both of these first two (or three) potential principles of energy justice could be used to guide future policy to ensure the most cost-effective energy choices are made and to ensure public interests are not sacrificed in order to advance private financial gains. While these principles of energy justice were derived from analyses of the arguments against nuclear power and the associated problems with this form of energy production, they also apply more generally to other potentially problematic energy systems. For example, some policymakers, observers, and analysis (especially those with ties to the oil/gas/coal industries) are strong proponents of so-called clean coal technologies, including carbon capture and sequestration (CCS) technologies. It is said this technology would facilitate the continued use of coal for energy production while significantly reducing the amount of greenhouse gases released into the atmosphere. The problem is, following from the first principle of energy justice above, these technologies do not yet exist anywhere in the world at the necessary scale. It would take *significant* amounts of time and money to make these technologies a reality, *if* such an undertaking is even possible. All the while the ultimate benefits/results will remain uncertain due to the fact that these currently remain hypothetical technologies. Especially when decisions about energy are being made in the context of efforts to address climate change, time and money are both in short supply.

The implementation of renewable energy systems based upon solar and wind production and energy efficiencies does follow from these first two principles of energy justice because such systems do not require massive forms of public financing, and the public financing of such systems that does exist (i.e., tax credits for new installed capacity) primarily benefits public as opposed to private interests. This is not to suggest the financing of renewable energy systems will be trivial. As it stands, the world currently spends about $1.8 trillion per year for all energy investments, which is close to 2 percent of world GDP. About half of that total (roughly $900 billion) goes to perpetuating the extraction and use of fossil fuels. Advocates for nuclear power and renewables scramble and compete for a share of the remainder. The question becomes, how governments and the public sector can steer most new investments into renewable energy and energy efficiency while keeping new investments in fossil fuels to a minimum that declines quickly over a short time? Furthermore, unlike nuclear energy, renewable energy is increasingly able to attract private investments. Thus, renewable energy is not totally dependent on public financing or prone to these distributive injustices.

Other, similar, issues limit the role nuclear power can possibly play in addressing climate change, beyond the prohibitive costs and lack of private investment. James Hanson, perhaps the best known and most respected climatologist in the United States, published a letter in the *New York Times* in 2013 where he argued the planetary threats posed by increasing greenhouse gas emissions are so grave that any and all efforts to curb their release must be undertaken. In this letter, Hanson singles out nuclear power as an "essential" component for curbing greenhouse gas emissions, citing the potential for "fourth generation" reactors, such as "pebble-bed reactors" and "thorium-fuel reactors" that are said to have solved all the problems of earlier generations. Hanson even went so far as to encourage people not to support antinuclear organizations due to their opposition to the development and implementation of nuclear technologies. Hanson is without question one of the foremost authorities on the issue of climate change. However, Hanson does not have a good understanding of the promises versus the realities of nuclear power. For one thing, as with CCS technologies, so-called fourth-generation reactors do not exist anywhere in the world, except on paper. Given that nuclear power companies have such difficulties trying to finance the construction of "conventional" reactors, none of them are racing to start developing new prototype reactors. There is no reason to believe that developing and constructing never-before-built "fourth generation" reactors will be any less prone to tremendous cost overruns and construction delays. Additionally, in order for nuclear power generation to make a significant contribution to curbing greenhouse gas emissions, several hundred new reactors would need to be constructed and brought online over the next ten to twenty years. Bracketing the issues of increasing the chances of nuclear disaster and the increase in the radioactive waste crisis that would follow from this approach (both of which are discussed below), there are only a handful of facilities in the world that can produce the massive, steel "reactor pressure vessels" that are needed to safely contain the nuclear reaction inside the reactors. According to studies by MIT and the International Atomic Energy Association (IAEA), available forges in Japan, China, Russia, and the United States together are only capable of producing about fifteen of these in a given year. Therefore, it is virtually impossible for enough new nuclear reactors to be constructed in a short enough time span to make any kind of a significant contribution to efforts to fight climate change.

Following the analyses of how antinuclear organizations use discourses of justice in their arguments against federal "loan guarantees" the investigation turned to antinuclear organizations' claims regarding the management of high-level radioactive waste in the United States. The high-level radioactive waste that is the unavoidable by-product of nuclear power production has been seen as the central issue with regard to nuclear power since the beginning of commercial nuclear power production, by both proponents and opponents of the technology (Walker, 2009). Additionally, the issues of radioactive waste storage and disposal are perhaps the most direct ways to see the past, present, and future environmental justice problems created by the production of nuclear power. The preceding analysis shows how the "unsurpassable social and political obstacles" that necessitated the canceling of the Yucca Mountain Project relate back to interrelated distributional, recognitional, and representative injustices that have plagued the radioactive waste management program in the United States since the very beginning.

Following from the analysis of antinuclear groups' claims on this issue, combined with analyses of the historical social, political, and economic contexts sorounding them, three additional possible principles of energy justice were established. The third principle of energy justice is: *energy justice requires that energy systems be developed in such a way as to minimize the production of hazardous by-products associated with energy systems to the greatest extent possible.* This would work to minimize the socially and environmentally harmful consequences of energy systems, which have historically disproportionately burdened minority and low-income communities. As Bullard (2000) clearly lays-out in one of the first sociological studies of environmental racism, whenever there are environmental benefits to be enjoyed and environmental burdens to be avoided, those benefits and burdens are distributed unequally along social hierarchies of race and class. This results in dirty, dangerous, and otherwise harmful facilities and by-products of energy production becoming concentrated in communities-of-color and other low-income communities.

In this case, what is true for environmentally harmful facilities in general is also true when it comes to the harmful effects of energy production. Sze (2005) clearly shows how the harmful effects of coal-fired power plants (air pollution, coal ash, dust, etc.) disproportionately and unjustly impact communities-of-color in New York City due to the fact that such plants are overly concentrated in these communities. Furthermore, as the analysis here has shown, efforts to manage the lethal by-product of nuclear power production (high-level radioactive waste) have consistently been at the expense of low-income, minority communities, especially American Indian communities in the Southwestern United States. From the operations of Nuclear Waste Negotiator and the search for a "monitored retrievable storage" site, to the railroading of the Yucca Mountain Project, to future efforts to find "volunteer host communities" for a nuclear waste repository, American Indian communities have consistently been targeted in proposals to manage radioactive waste. Therefore, policy decisions following this third principle of energy justice would help eliminate this problem at the source, by minimizing (to the greatest extent possible) the production of hazardous materials associated with energy production that would otherwise most likely become concentrated in poor and/or minority communities. Clearly, following from this third principle of energy justice, nuclear power should no longer be pursued because its production is based upon and reinforces multidimensional social and environmental injustices.

While the hazardous by-product of nuclear power is somewhat unique in its sheer-concentrated-lethality and the length of time it will remain a hazard, this principle of energy justice also applies to other dirty and dangerous methods of energy production. For example, it could be argued that the growing release of CO_2 and other greenhouse gases from coal-fired power plants constitutes a global hazard that following the third principle of energy justice should be reduced to the greatest extent possible. Additionally, following from this principle, the hazards posed by coal ash, as well as the hazards and devastation caused by mountaintop removal mining, also mean that energy systems based upon coal should be replaced with other cleaner and safer forms of energy production (such as solar and wind, which are discussed below.)

The fourth principle of energy justice (the second following from the analysis of the radioactive waste issue) is: *energy justice requires the explicit acknowledgment of preexisting injustices and inequalities combined with concrete policy mechanisms to ensure the burdens are not being disproportionately bore by the most vulnerable.* The analyses of past, present, and (potential) future environmental injustices associated with efforts to manage high-level radioactive waste, which clearly showed how previous multidimensional injustices and preexisting inequalities impacted present processes, led to the proposal of this principle highlighting the role of history in current conflicts. If energy justice is the goal, then present and future decisions about the development and implementation of energy systems must be made by acknowledging the existence of past and current social inequalities, in order to avoid reinforcing these inequalities through the development of energy systems. However, to achieve energy justice, it is necessary to go beyond simply acknowledging social injustices and inequalities in decision-making and political processes. This explicit acknowledgment must be combined with inflexible policy mechanisms and procedures intended to ensure the harmful effects of a given energy system are not being unjustly concentrated in poor and/or minority communities. This could include the establishment of meaningful/powerful oversight and regulation of energy decisions that are independent of both the energy industry and their allies in the state; perhaps through the empowerment of citizens councils established to ensure environmental justice requirements and directives are enforced. Of course, decisions made following the third principle of energy justice (above) will render the fourth principle increasingly unnecessary, as safer and cleaner energy systems are pursued that produce less and less hazardous by-products that might otherwise be concentrated in marginalized communities.

The fifth principle of energy justice (the third and final principle to emerge from the analyses of the radioactive waste issue) is: *energy justice requires the establishment of preexisting guidelines and standards meant to protect public health and the environment, which are less open to manipulation than site-specific guidelines.* This would not only greatly increase public and environmental protections from the human health and environmental dangers posed by some energy systems, but would also greatly increase public confidence in both the energy industry and especially in regulatory bodies meant to oversee the industry. The analysis of the antinuclear movement's claims regarding the multidimensional injustices associated with both the Yucca Mountain Project and the ordinary operations of the Nuclear Regulatory Commission (NRC) led to the identification of this principle. In both cases, "site-specific" guidelines, as opposed to "general" guidelines, were easily and repeatedly (in the case of the NRC) relaxed in response to the failure of specific sites to meet those guidelines. "General" guidelines that are developed independently of considerations of any one site or the interests of specific nuclear operators (or other specific economic actors) have been shown to be less open to this kind of manipulation. Therefore, with the promotion of social and environmental justice associated with energy systems as the goal, energy justice requires that general guidelines be put in place to guide the development of energy systems in such a way as to ensure the protection of public health and the environment is not sacrificed in order to advance certain project or private interests.

Turning attention from analyses of the claims made by national level antinuclear organizations regarding federal "loan guarantees" and the management of radioactive waste to the claims made by local level antinuclear organizations pertaining to CWIP financing and campaigns to increase the safety of nuclear facilities allows for the identification of three additional principles of energy justice. The principles of energy justice that follow from the analyses of local level organizations' claims in these areas share certain similarities with those above, as well as complement and supplement them. The sixth principle of energy justice (the first following from the analyses of local level antinuclear organizations) is: *energy justice requires that decisions about the development and implementation of energy systems need to include mechanisms for facilitating public involvement at all relevant levels to the greatest extent possible.* Because decisions about energy systems in the United States are made on multiple levels (local, state, regional, national) and because the impacts of those decisions are felt on multiple levels, democratic decision-making based upon public input and participation is required on all these levels so that the public interests on any one of these levels are not sacrificed for those on another (or for private interests of capital accumulation).

T... principle of energy justice emerged from reflections on the analyses ...aims made on both the national and local levels of the antinuclear ...t in the United States. While it is true that most of the politics and ...bout nuclear power in the United States are made at the national level, important aspects of the multidimensional and interrelated injustices associated with nuclear power (and other dirty, dangerous, and expensive forms of energy production) are revealed by looking to local level conflicts and contexts. Granted, it will be exceedingly difficult, if not impossible, to ever fully implement this kind of multilevel democratic decision-making procedure in the area of energy policy (or any other policy arena). Time is in short supply with regard to the onset of the effects of climate change, and this method of deep-democratic decisions making surely will not be conducive to making quick decisions. However, it remains a desirable goal to strive toward, even if it is never fully attained. The past decisions that have resulted in our current fossil fuel and nuclear-based energy system were not made even close to democratically, and, at least partially as a result of this, have been shown to reproduce social and environmental injustices and inequalities. The establishment of socially, politically, economically, and environmentally just energy systems (i.e., energy justice) will therefore require the establishment of deeply democratic methods of decision-making on multiple levels.

The seventh principle of energy justice is closely associated with the third principle outlined above. More specifically, *energy justice requires energy systems be developed in such a way as to reduce the possibility of catastrophic accident, even so-called low-probability, high-consequence catastrophes, to the greatest extent possible*. This principle follows from the analysis of antinuclear groups' claims in conflicts over increasing the safety of nuclear facilities in the post-Fukushima context, and is in fact one of *the central* lessons that should be derived from Fukushima and other energy disasters (Three Mile Island, Chernobyl, BP's Deep-Water Horizon blowout, etc.). While the consequences of energy disasters are truly global in scope (and in the case of nuclear meltdowns, they are disasters with a beginning but practically *no end*), the most severe and immediate consequences are felt locally by marginalized communities living proximate to the source. Thus, the analyses of the discourses of justice used by local level antinuclear organizations, combined with analyses of the relevant contexts, supplement and support the first principle proposed earlier, which followed from analyses of national-level organizations. This confirms Fraser's (2009) contention that "issues of scale" are important when exploring existing injustices and/or theorizing just alternatives. In short, the magnitude of the potentially catastrophic consequences that can result from a nuclear accident means, following this principle of energy justice, nuclear power should not be pursued moving forward. Safe and clean energy systems are increasingly becoming a possibility, and energy justice requires that they be developed to the greatest extent possible.

The final principle of energy justice that is revealed through analyses of the antinuclear movements' claims and the associated contexts is related in many respects to some of the previous principles. That is, the eighth principle of energy justice (the third derived from the analyses of local antinuclear organizations, and the final principle to be proposed in this study) is: *energy justice requires that the implementation, production, and distribution of energy systems must include strong, effective, and independent regulatory oversight.* This means energy justice requires concrete mechanisms to ensure "regulatory capture" does not result from the industry to be regulated effectively taking control of the regulatory body itself. There are fairly obvious, deep, structural problems with asking agencies of the state to effectively regulate private industries (in order to protect public health, well-being, and the environment) when the agency is itself financially dependent on these same private industries for its continued functioning. Furthermore, recent history has provided ample examples of the "revolving door" between industry officials, insiders, and lobbyists, on the one hand, and supposedly independent regulatory agencies, on the other.

Although the necessity of impactful oversight has been implied in some of the previous principles discussed, this principle is included as a standalone because of the NRC's unwillingness and/or inability to effectively and adequately regulate the nuclear industry in the service of protecting public interests, public health, and the environment. Possible mechanisms to stop the "revolving door" between industry insiders, lobbyists, and regulatory bodies could include potential third-party audit/review of the regulator's record with regard to public and private interests. Additionally, meaningful public input into the appointment and confirmation processes for regulators could function to break up the incestuous relationships between regulators and regulated industries. The bottom-line is decisions about the implementation and development of energy systems are too important to the functioning of twenty-first-century societies (not to mention the global climate), and the consequences of these decisions have far too great an impact on too many people to be inadequately regulated. The history and current practices of regulation over the energy industry, including the nuclear power industry, show time and time again the financial interests of the industry are often prioritized at the expense of protecting the public interest, public health, and the environment.

As has already been said, the goal in establishing this framework for energy justice is to take steps toward establishing sociologically informed arguments in opposition or support of various choices about the development and implementation of energy systems. These principles are by no means definitive, nor necessarily complete. Further research into the discourses of justice utilized by environmental justice energy activists (including antinuclear activists) and their claims against dirty, dangerous, and expensive energy systems, combined with analyses of the relevant historical and social, political, and economic contexts, will surely reveal additional dimensions of injustice associated with current energy systems. This will assuredly facilitate the identification of additional principles of energy justice not included here.

Considering that these principles, and the framework they collectively provide, were developed through analyses of the claims made by antinuclear organizations, it should not be surprising that together these principles suggest nuclear power should no longer be pursued due to the multidimensional and interrelated injustices involved. For one thing, nuclear power is prohibitively expensive. The tremendous up-front costs and the extremely long construction schedules, combined with the high risk of eventual project failure (~50 percent), means that every nuclear development project in the United States and around the world is being publicly financed, due to a complete lack of private investment. And at least in the United States, despite the fact that nuclear facilities are publicly financed, the eventual/potential revenues and profits are retained by privately owned nuclear utilities and their investors. Additionally, nuclear power is unique among energy sources in the extraordinary hazardous and lethal nature of the waste it produces, and the severity and extent of potential accidents. No other energy source produces such extraordinarily lethal and long-lived by-products that require total isolation from all living things for hundreds of millennia. No other energy source creates the same possibility of large areas of the Earth becoming forever uninhabitable due to severe radioactive contamination. Given these facts, no one can *honestly* claim that nuclear power is a "clean" energy source that belongs in the same conversation as other clean, renewable energy sources like solar and wind power. Furthermore, and just as significantly, the historical and current operations of nuclear power production in the United States have been/are facilitated by undemocratic decision-making that fails to adequately regulate the commercial nuclear industry and continually advances the private interests of the industry at the expense of the public interests of communities, taxpayers, and ratepayers. The production of nuclear power has in part been made possible by exploiting and reinforcing existing social injustices and inequalities, including a lack of democratic decision-making involving affected communities and individuals on different levels. This often involves the use of "site-specific" guidelines that can more easily be altered to reflect private economic interests, rather than "generic" standards that have been shown to better reflect and protect the interest of public health and the environment.

Each of these problems with nuclear power indicates that from a sociological and social justice point of view, nuclear power is not an acceptable option for current or future energy systems. The argument against nuclear power is especially strong when all of these problems are taken together and evaluated through an energy justice lens. However, as suggested earlier, these principles of energy justice can also be used to make arguments for and against other types of energy systems currently being used and considered around the

United States and around the world. For example, these principles could be used to construct a sociologically informed argument against the continued use of coal, natural gas, and/or petroleum. Especially on a global scale, the increasing release of CO_2 and other greenhouse gases from coal-fired power plants surely constitutes a hazardous by-product, not to mention the coal ash ponds, the slurry, and the extensive environmental damages caused by mountain top removal. Energy justice requires that these types of hazards be minimized to the greatest extent possible. Similarly, issues surrounding shale gas extraction (or natural gas "fracking") have recently caused significant conflict in parts of the United States. The lack of adequate regulation of the natural gas industry and a lack of democratic decision-making on relevant levels have resulted in "fracking" operations being exempted from the Clean Water Act and not having to report the ingredients of the fluids they use in their operations. Energy justice would require strong and meaningful regulation of the natural gas industry, occasioned by strong democratic decision-making on the local, regional, and national levels. Additionally, deep-sea oil extraction, Arctic drilling, and Tar Sands oil extraction and transport pipelines all create the possibility for extreme catastrophic accidents. Given that other technologies and potential energy systems exist that do not carry these kinds of catastrophic environmental and social risk, energy justice requires that the safest cleanest technologies be developed to the greatest extent possible.

The principles of energy justice established here are not specific to the case of nuclear power, but rather can be used to evaluate different types of energy systems and technologies. Furthermore, these principles of energy justice can be used to make sociologically informed arguments *in favor* of the development of clean, safe, renewable, and affordable energy systems, such as those based upon wind, solar, and energy efficiency. However, before developing this argument in favor of clean and renewable energy systems, it is necessary to first reflect upon the relationships between the different dimensions of the interconnected injustices associated with nuclear power production, which have emerged from the various analyses above.

COMMON PROGRESSION OF THE DIMENSIONS OF INJUSTICE

To different extents in the cases discussed earlier, it seems that within conflicts over dirty, dangerous, and expensive energy technologies, including nuclear power, representational injustices often lead to and facilitate recognitional injustices, which together reinforce and recreate distributional injustices of environmental and economic benefits and burdens. Put another way, it is often the case in conflicts over nuclear power that representative injustices, embodied in biased/undemocratic decision-making processes and procedures, lead to the institutionalized misrecognition of public interests and desires on different levels, which furthers distributive injustices, both in terms of economic and environmental benefits and burdens, and in terms of public versus private interests. This basic progression of different dimensions of injustice is represented in figure 4.1.

However, this simple, linear progression imposes a somewhat artificial starting point, because the original representational injustices emerge as a result of other preexisting recognitional and distributive injustices (represented by the bottom arrows running right to left). While this model is by no means definitive, it does help illuminate the possible relationships between the multidimensional and interrelated injustices associated with nuclear power and other dirty, dangerous, and expensive energy systems. Furthermore, this model helps us think about the deep connections between social, environmental, and energy justice, on the one hand, and democracy and democratic decision-making on the other.

This progression or flow of interrelated injustices emerges to different extents in the various conflicts over nuclear power discussed above. For example, in the conflicts over the possible expansion of federal "loan guarantees" for new nuclear reactor construction, this plays out on the national level with regard to financing new nuclear facilities. The representational injustices in this case were manifested in institutionalized bias in the decision-making process at the congressional level. Key members of Appropriation Committees and Water and Energy Committees, who were working to advance legislation to expand the "loan guarantee" program, were not representing the interests of citizens and taxpayers. Rather, due to the influence granted to the nuclear industry through their lobbying efforts and campaign contributions, the democratic process was being subverted through the representation of industry interests over public interests. This is closely connected to the resulting misrecognition of public concerns and interests and the eventual maldistribution of economic benefits and burdens.

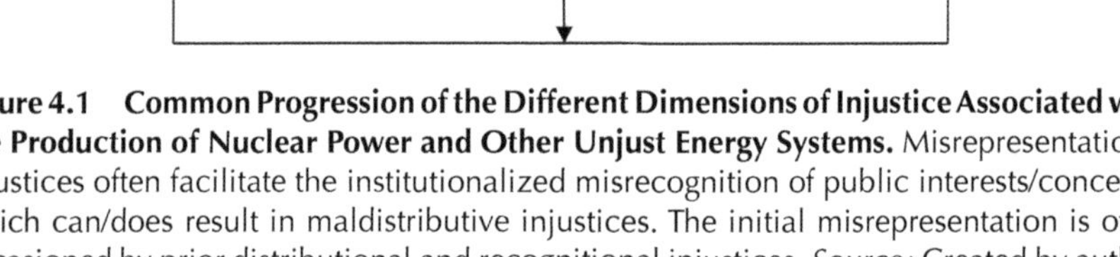

Figure 4.1 Common Progression of the Different Dimensions of Injustice Associated with the Production of Nuclear Power and Other Unjust Energy Systems. Misrepresentational injustices often facilitate the institutionalized misrecognition of public interests/concerns, which can/does result in maldistributive injustices. The initial misrepresentation is often occasioned by prior distributional and recognitional injustices. *Source:* Created by author.

More specifically, following the representational injustices, public concerns about the risks taxpayers would bear by financing new reactor construction were not justly recognized within the decision-making process, rendering these concerns practically invisible. And with the misrecognition of public concerns in place, the maldistribution of economic benefits and burdens between private and public interests would have ensued as the program moved forward.

A very similar progression of the different dimensions of injustice also emerges from the above analysis of local antinuclear group's campaigns against CWIP charges. In this case and on this level the initial representational injustices emerge within the institutionalized bias in decision-making procedures at the state and/or local level, including state legislatures, the operations of Public Service Commissions, Public Councils, and other local-level regulatory bodies. These representational injustices involved limiting democratic decision-making and public oversight over the development of energy systems by essentially granting electric utilities a blank check to proceed with expensive and financially risky development projects. Thus both CWIP financing and federal "loan guarantees" are mechanisms for socializing the risk of nuclear energy development projects, while allowing private, investor-owned corporations to retain control of any eventual benefit/profit. Consequently, CWIP financing is advanced through the subversion of the democratic process that is occasioned by the undue influence the nuclear industry is able to exert through its financial resources and lobbying efforts. As before, but on the local level, this unjust misrepresentation of public interests and desires feeds directly into the institutionalized misrecognition of these same public desires and interests. Public concerns and fears regarding the unjust risk of public financing for new reactor construction remain misrecognized in the political processes, which paves the way for the maldistribution of benefits and burdens between public and private interests. The inequitable maldistributive injustice in this case is especially evident with regard to people who will ultimately be paying for a product and/or service they will never receive.

While this pattern of misrepresentation leading to misrecognition leading to maldistribution (with the initial misrecognition in part resulting from previous recognitional and distributive injustices) can be seen in the cases of conflict over the public financing of new nuclear reactors, it is especially evident in cases of conflict over increasing the safety of nuclear facilities and the management of radioactive waste. As shown in the above analysis of local antinuclear organizations' efforts to increase the safety of nuclear facilities, the operation of unsafe nuclear facilities is based upon representational injustices within the operations of the NRC and state-level regulators that prevent sufficient public participation in decisions about the (continued) operations of these nuclear facilities. The analysis of the claims made by local antinuclear organizations shows how these representational injustices lead to the unjust misrecognition of public concerns and interests, which ultimately facilitates the maldistribution of benefits and burdens between public interests (local communities, citizens/taxpayers, consumers/ratepayers) and private interests of capital accumulation (the nuclear industry including investor-owned utilities). More specifically, the analysis of the campaign to close Vermont Yankee especially highlights the central problems with misrepresentation of public concerns/desires within the NRC leading to the misrecognition of these and ultimately the maldistribution of both social and environmental benefits and burdens. As the analysis of these local antinuclear groups' claims brought to light, democracy was not just being *passively* subverted in this case; it was being *actively* subverted by Entergy, a "mega-corporation" who along with the rest of the nuclear industry use their tremendous resources to gain undue influence over the political and regulatory process. Therefore, the concerns of the people in this case were institutionally misrecognized at the federal level by NRC, which enabled the distributive injustice of radioactive waste to continue to accumulate on the banks of the Connecticut River.

While the progression of the different dimensions of injustice proposed here is fairly clear in this case, this same progression also clearly emerges in different ways at different points in time and on different levels in the case of past and current efforts to manage high-level radioactive waste. There has been a clear historical, national pattern of targeting Native American lands and communities with radioactive waste dumps. As the above analysis has shown, the brief operations of the Office of the Nuclear Waste Negotiator consisted primarily of trying to exploit past and current representational injustices within Tribal governments and decision-making procedures, which involved the deep misrecognition of Tribal members' fears and concerns (in some cases a total lack of recognition, rendering people's concerns functionally invisible). And of course, this was all part of an effort to establish and reinforce distributive injustice by concentrating radioactive waste on Tribal lands, which would clearly harm/burden Tribal members far more than any other section of the public. As Kamps (2005) argued,

> The economic impoverishment of tribes, the tendency for tribal leaders to act without the authorization or even knowledge of their people, and the Bureau of Indian Affairs' ability to interfere in internal tribal affairs arbitrarily and capriciously . . . all mean that these siting processes were all too often not fair, nor truly voluntary.

This example surely highlights the ways that preexisting distributive and recognitional injustices set the stage for and facilitate representational injustices that in turn reinforce and recreate further recognitional and distributive injustices through time.

Almost twenty-five years later this same progression of the different dimensions of injustice emerges in the procedures and eventual recommendations of the BRC. As the analysis above demonstrates, before the BRC even held their first public meeting, it was clear the Commission was being built around representational injustices on at least two levels—what Fraser (2010) calls the political injustices of "misframing" and "ordinary misrepresentation." The former relating to the fact that *the central question* regarding whether or not to continue to pursue nuclear power was framed as being "outside the scope" of the BRC, and thus did not include public participation or even comment. The latter relating to the fact that the structure of the decision-making process functions wrongly to exclude those who (in this case) are most impacted by the decision reached regarding the handling of radioactive wastes (e.g., American Indian groups, reactor communities, uranium mining communities, etc.). However, the "nuclear-industrial-academic-complex" (Kuletz, 1998) was well represented among the Commission members. The majority of the Commission's members were former pronuclear legislators, former nuclear regulators with many long-standing ties to the industry, nuclear industry executives, and academic experts committed to nuclear technologies. This unjust misrepresentation clearly and explicitly (especially in the case of "orphaned" reactor communities) leads to the institutionalized misrecognition of affected communities and the public's concerns and interests. And furthermore, the BRC's primary recommendation to establish one or more "centralized interim storage" sites would by definition create distributive injustices for whomever (most likely American Indian groups) lives in proximity to these sites.

In short, reflecting upon the analyses performed here on the discourses of justice utilized and reconstructed in the claims made by antinuclear groups, as well as the analyses of the historical and social, political, and economic contexts from which they emerged, reveals a common progression of the different but interconnected dimensions of injustices that are associated with the production of nuclear power and other dirty, dangerous and expensive energy systems. It could be argued that perhaps efforts to identify all three dimensions of injustice (distribution, recognition, and representation) in these cases are stretching Fraser's (2003, 2009) model a bit far, especially with regard to the dimension of recognition being applied too loosely. There might be some validity to this argument, but it is necessary to point out that it is not being argued all three dimensions are always equally significant in all cases. It is true the representational and distributive injustices identified above are the most significant dimensions in the cases in question. However, it is also true that there are recognitional injustices involved in these processes, and these provide a conceptual and empirical link between the more obvious representational and distributive injustices. Fraser's (2009) central point in arguing for a three-dimensional conception of justice is real-world injustices are generally rooted in the social, political, and economic spheres of society, and thus just alternatives must also consider all three dimensions.

Additionally, part of this potential issue is Fraser's (2000, 2003) institutionalized status model that sees misrecognition as being connected to social structure and institutions. This is a valuable and important insight into the realities of misrecognition, vis-à-vis the more Hegelian, free-floating discourse model of recognition scholars such as Honneth (1996, 2012). Fraser's (2000, 2003) "status model" of recognition makes misrecognition easiest to identify in decision-making and/or political processes. Because what needs to be recognized is culture, identity, and other desirable differences *within institutionalized frameworks of decision-making*, as opposed to within free-floating discourse, issues of recognition and representation are necessarily closely connected within Fraser's model. While the issues of misrepresentation and misrecognition are closely connected in practice, it is a mistake to collapse these categories by only seeing the representative dimensions. What we have in the cases analyzed for the present study are closely related, yet distinct, representational, and recognitional injustices, in addition to issues of maldistribution.

Furthermore, one of the major sources of analytic utility that the model proposed here provides is it helps illuminate the deep and necessary connections between the advancement of social justice, on the one hand, and democratic governance and decision-making, on the other. More exactly, this model for conceptualizing the flow and/or progression of different dimensions of injustice in real-world struggles over energy gives issues of representation (and thus democracy and the political sphere) a very central role. Past and preexisting recognitional and distributive injustices feed into and help create representational injustices. But it is these representational injustices that facilitate the *institutionalized* misrecognition of communities', taxpayers', ratepayers', and generally the public's concerns and interests. Therefore, it is a closely connected combination of misrepresentation and misrecognition that gives rise to maldistributive policies concerning environmental and economic benefits and burdens.

The primary goal of "emancipatory social science" (Wright, 2012) should be to help keep fragile, democratic experiments alive and to help them flourish and grow. This is reflected in the fact that issues of democratic participation in decision-making and adequate public oversight and regulation are either directly or indirectly connected with each of the principles of energy justice proposed here. Utilizing these principles (and the framework they collectively produce) allows for the construction of arguments *against* the development and implementation of nuclear power and other dirty, dangerous, and expensive energy systems, based in large part on the fact that such developments are carried out through the subversion of democratic processes. However, these principles of energy justice and the democratic ideals that are either directly or indirectly connected to them can also be used to make a sociologically informed argument *in favor* of energy systems based upon renewable energy sources, such as wind and solar power. The development and implementation of such a renewable energy system can be done democratically, and in fact could be a part of strengthening democratic decision-making processes in the United States in general.

ETHICS AND INNOVATION

Questions of (in)justice inevitably bump up against questions of ethics—the ways people can tell right from wrong, or good from evil. Ethical principles are part of a normative theory that justifies or defends moral rules and/or moral judgments; they are not dependent on one's subjective viewpoints. Some fundamental ethical principles include respect for autonomy (also called the principle of human dignity), beneficence, nonmaleficence, and, of course, justice. Respect for autonomy means there is an obligation to respect the decisions made by other people concerning their own lives. It provides a negative duty not to interfere with the decisions of competent adults, and a positive duty to empower others. The principle of beneficence means there is an obligation to bring about good in our actions, and we must take positive steps to prevent harm. Similarly, the principle of nonmaleficence holds there is an obligation to not harm others, and when harm cannot be avoided, there is an obligation to minimize the harm done. Corollary principles to the principle of nonmaleficence include that it is wrong to increase the chances of harm to others, and it is wrong to waste resources that could be used for good.

The present study does not just simply stand pat with the identification and enunciation of principles of energy justice. On the contrary, the present study has an ethical component based upon the argument that the principles of energy justice *should* be utilized in decisions regarding energy systems and technologies. This is based on an implicit conditional statement. *If* justice is a goal of energy policy, *then* the principles of energy justice should be used for making relevant decisions. Furthermore, ethical principles guiding what should be done with regard to energy policy decisions imply what can be done—should implies can. It is not a valid ethical argument to insist energy policies and technologies be developed with no limitations, costs, or burdens. This cannot be done. All energy technologies and energy systems have limitations and shortcomings, including renewable energy systems based upon wind and solar power (Perkins, 2017). However, it is possible to make ethical and effective decisions about energy technologies and systems that maximize energy justice.

Ethical decision-making refers to the process of evaluating and choosing among alternatives in a manner consistent with ethical principles (i.e., justice). In making ethical decisions, it is necessary to perceive and eliminate unethical options and select the best ethical alternative. Good decisions are both ethical and effective. Ethical decisions generate and sustain trust, demonstrate respect, responsibility, fairness, and caring, and are consistent with good citizenship. These behaviors provide a foundation for making better decisions by setting the ground rules for our behavior (Josephson Institute of Ethics, 2002). Decisions are effective if they accomplish the intended goals and purposes. A choice that produces unintended and undesirable results is ineffective. The key to making effective decisions is to think about choices in terms of their ability to accomplish our most important goals. This means we have to understand the difference between immediate and short-term goals and longer-range goals.

The importance of (and connections between) ethical and effective decision-making with regard to nuclear power can easily be seen when comparing the approaches to high-level radioactive waste storage in Finland and the United States. Finland is the only nation actively moving forward with the development of a geological repository to permanently store the high-level radioactive waste produced from their commercial nuclear power reactors. If all goes according to plan, sometime within the next ten years the first of what will eventually be nearly 3,000 sealed copper canisters, each up to 17 feet long and containing about 2 tons of spent reactor fuel from Finland's nuclear power industry, will be lowered into a vertical borehole in a side tunnel about 1,400 feet underground. As the canisters are buried, the holes and up to 20 miles of tunnels will be packed with clay and eventually abandoned. Between the engineered barriers (such as the 2-inch thick copper canisters), the clay, and the surrounding ancient granite formation, the Onkalo repository is expected to safely seal the irradiated fuel rods for tens of thousands of years. The repository is expected to take about a century to fill with a projected cost of nearly $3.9 billion and will be the world's first permanent disposal site for high-level radioactive waste.

It is easy to marvel at the technological achievements of the Onkalo repository, but what is most amazing is the project has been proceeding smoothly for years with the support of the local population and municipal government. As discussed above, this stands in marked contrast to efforts to construct a similar facility at Yucca Mountain, Nevada, which has faced political opposition from environmental and antinuclear groups, citizens, and Nevada lawmakers for years. The success of the Finnish project is due in part to how it was presented to the people who would be most affected by it. Each community under consideration as a repository location was consulted and promised veto power should it be selected. Those in charge of the Finnish site selection process (beginning in the 1980s) learned early lessons about the need to consult with local residents. Initially, premade industry decisions were announced as unreviewable, which inevitably lead to local opposition and push back. However, officials quickly learned they had to be very open and transparent, which helped to create and facilitate trust. The approach proved so successful that when it came time for the national government to make a final decision on a repository in 2000, officials in the local municipality agreed to host it on one condition: that Posiva (the company building the repository) not present the government an option to choose any other site. The technological achievements of Onkalo are equaled (or exceeded) by the social and political achievements.

In contrast, in the United States, Congress in 1987 pre-emptively directed that only Yucca Mountain be studied as a potential site, effectively overruling opponents in Nevada who were worried that the project might affect water supplies or otherwise contaminate the region. Unlike officials in Finland, officials with the DOE tasked with constructing and maintaining a high-level radioactive waste repository in the United States have failed to move beyond the "decide-announce-defend" model of public engagement. The project at Yucca Mountain, in the Mojave Desert about 100 miles northwest of Las Vegas, has been studied for years at a cost of more than $13 billion. After decades of "site suitability studies" and tens of billions of dollars, the Yucca Mountain facility consists of little more than an exploratory tunnel. Furthermore, members of Nevada's congressional delegation are still vowing to fight the project should it be resuscitated, arguing that there are concerns about the long-term safety of drinking water supplies (unlike the Finnish repository, the Nevada site sits above the water table) and that, above all, Nevadans do not want it.

This comparison clearly demonstrates the importance of ethical decision-making based upon ethical principles (such as justice) for achieving effective outcomes. There are practical, pragmatic reasons for establishing energy justice, in addition to the more obvious moral reasons. The Finnish program is moving forward because the officials in charge of radioactive waste disposal established just representation and recognition of local stakeholders. Transparency and meaningful participation in the decision-making process established trust, which minimized local pushback and opposition. However, DOE officials in the United States have misrepresented local stakeholders in the decision-making process and misrecognized their views from the outset of the program. This has predictably resulted in consistent public opposition to the Yucca Mountain facility and is the primary driver of the unsurpassable technical and *social* obstacles that forced the abandonment of the project in 2008.

POTENTIAL CONCEPTUAL AND METHODOLOGICAL SHORTCOMINGS

The conceptual frameworks utilized and developed for this study, as well as the methodological approaches taken, are especially well suited for addressing the central research questions that motivated this project. However, as with all theoretically informed empirical research projects, specific conceptual and methodological decisions produce both advantages and disadvantages. And because no one study, no one conceptual framework, nor one research methodology can do everything, certain possible critiques and/or shortcomings of the present study emerge. For instance, it could reasonably be asked, "Why do we need energy justice" and "how is it different from environmental justice." These are fair questions considering the close connection between energy and environmental issues and injustices. However, it is necessary to think about energy conflicts as distinct because of the specific, vital role energy systems play in the development and functioning of modern societies (Nye, 1998). Decisions about the development and implementation of energy systems have significant consequences for all social, political, and economic organizations, including the mollification or enhancement of social inequalities and hierarchies. As new energy systems are developed and implemented, it significantly alters exiting forms of production, consumption, and communication. The far-reaching social, political, and economic impacts of energy system development and implementation mean special attention must be paid to the connections between these systems and the development or hindrance of social justice. Furthermore, energy issues (including issues of energy justice) can be accurately conceptualized as a subcategory of broader environmental issues. The relative benefits and burdens of different energy systems can be conceptualized in environmental terms, relating to environmentally just or unjust arrangements. However, it is important to work toward energy justice *as well as* environmental justice because focusing on the specific goals of energy justice will contribute to the achievement of the broader goals of environmental justice.

A second related possible critique has to do with the use of the liberal political categories of distribution, recognition, and representation that are central to the present study. For some social scientists, especially those who have taken the cultural/discursive turn, the use of these (or frankly any other) categories is highly suspect and could be seen as an artificial imposition of preconceived categories on to the social world. The concern from this point of view could be that the use of these categories does not advance understandings of the issues involved, but rather simply substitutes the researcher's views of the issues for those of the subjects of this study. In short, the concern may be that this approach essentializes and reifies the complexity of the data.

This possible critique reflects a very deep division within the social sciences generally, and the discipline of sociology specifically. Following the downfall of historical materialism's paradigmatic dominance in the social sciences, many researchers and theorists began rejecting theories and conceptual frameworks that emphasized commonalities and "sameness" in favor of those that emphasized complexity, contingency, and "difference." This rejection was occasioned by the correct realization that some (perhaps many) uses of historical materialism resulted in overly deterministic views of the social world, which left little to no room for variations, contingency, complexity, and/or human agency. The constructivist turn resulted in tremendous increases in our understandings of how symbolically mediated communication and interaction resulted in the social construction of reality. The problem is, however, that during this paradigmatic transition some researchers have "thrown the baby out with the bathwater," and effectively swung the conceptual pendulum to other extreme by failing to recognize and/or acknowledge commonalities among or between social groups. This is exactly what Fraser (2003) refers to as the "problem of displacement," wherein theories of recognition do not so much complement/supplement theories of distribution, but rather replace them all together.

It is not a coincidence that this "problem of displacement" has emerged in *exactly* the same historical period that has seen the global rise of neoliberal economic theory that has drastically accelerated domestic and global economic inequalities to levels never seen before. In other words, at the moment in time when what is needed are strong and robust theories of distributive justice to combat accelerating inequalities associated with neoliberalization, what is happening is the displacement of these theories with theories focusing solely on recognition and difference. This is precisely why the present study utilizes and builds upon Fraser's framework, because it enables the bringing together (into a single framework) concerns with the discursive construction of reality and historical material conditions. While it is true that it is easier and "neater" to develop social theory from one extreme or the other, theoretical extremism is not useful, whether it comes from the historical materialist side or the social constructivist side. The social world is extremely complex. It follows that theories of the social world should reflect that complexity and resist any and all urges to present the social world as something that can only be understood in this or that way. This is why Fraser's framework, and the liberal political categories it builds upon, was an appropriate framework for the present study. Real-world injustice, including those associated with the energy systems, simultaneously stem from the social, political, and economic spheres of society. Thus we need a theoretical framework that is sensitive to not only historical material political and economic structures, but also to discursive constructions in the realm of cultural interaction and communication. And

Fraser's concepts of distribution, recognition, and representation allow for the simultaneous consideration of both these domains, as well as a synthesis of the insights that are generated from both ends of the conceptual spectrum.

A third possible area of critique of the present study has to do with the methodological decisions made and the approaches utilized. More specifically, it could be seen as problematic that the data is limited to claims made by antinuclear groups during the eighteen-month timeframe from January 1, 2010, through August 1, 2011. This decision creates the possibility that the conclusions reached are specific to this particular period of time, and are not generalizable beyond this eighteen-month window. However, the decision to focus on this particular period of time was made for strategic and theoretical reasons. This eighteen-month timeframe was selected as the basis for the analysis because it encapsulates many important developments in nuclear energy policy in the United States, which resulted in large volumes of social movement activity and discourse. More specifically, this period of time extends from the cancelation of the Yucca Mountain Project and the inception of the BRC (March 2010), through the release of the BRC's final report (July 2011). This eighteen-month timeframe also includes the movement's activities and claims made during the first six months of the ongoing Fukushima nuclear disaster, which allows for an analysis of how this event impacted antinuclear social movement activity. It is true that this eighteen-month period is not representative of the "typical" amount or type of antinuclear social movement activity in the United States. There was far more activity in this period than in any other eighteen-month period that preceded it over the previous decade. However, for this reason, this period of time is especially illustrative of the breadth of antinuclear social movement activity and examining these activities in this period allowed for the formation of general conclusions about the movement on different analytic levels. Furthermore, the historical materials included in the data and discussed in the analyses indicate the movement's uses of discourses of justice, and the interrelated and multidimensional injustices of nuclear power these reflect, are not unique to this period of time.

A possible related methodological critique has to do with the decision to focus on organizational claims-making in political conflict as the unit of analysis for this study. The study focuses on the ways antinuclear social movement *organizations* behave and act as political, organizational actors in the course of political conflict with other political/organizational actors (i.e., the state, regulatory agencies, the nuclear industry, utilities, etc.). Of interest was how antinuclear *organizations* utilized and reconstructed different paradigms of social justice in their claims-making in political conflict with other organizational, political actors. Thus, SMOs operating on different levels are my unit of analysis, with a focus on organizational discourse as it relates to broader, cultural discourses on energy policy, democracy, and justice. And because of the intensive focus on organizations' and the movement's textual materials, textual/discourse analysis is especially well suited for addressing "how questions"—how does the movement express its grievances; how does the movement construct and utilize frames and conceptualizations of justice; how does the movement utilize broader cultural symbols and representations.

However, by focusing the analyses on the level of organizational claims-making during the political conflict there are important questions that are not possible to consider. For example, the data does not reveal anything about the personal/individual drives, motivations, and desires that fuel activists' social movement participation. Nor does the data reveal how perceptions and experiences with broader systems of racism, sexism, class domination, and so on propel people to dedicate their time and energies to fighting nuclear power. Similarly, the data and methodological approach do not reveal the roles of race, class, gender, and so on play within antinuclear organizations, which could significantly impact the organization's claims themselves, and thus alter the conceptualizations of energy justice. Relatedly, the present study focuses exclusively upon *public* claims-making in the course of *political conflict*. This means that the claims being made might reflect what is perceived to be the most strategically advantageous form of argumentation, rather than what are perceived to be the biggest or most serious problems with nuclear energy and other dirty, dangerous, and expensive energy systems. However, following Somers' (2008) point about the normative and the empirical being mutually interdependent, there is analytic utility in looking at exactly how claims are made to maximize strategic advantage in different contexts as a means for identifying the structural and normative features of both just and unjust social arrangements. Focusing on the discourse used in claims-making reveals the normative component of (in)justice reflecting beliefs and ideas that themselves have causal powers. Focusing on the historical and social, political, and economic contexts that make different discourses strategically advantageous in different settings reveals the empirical and structural components of (in) justice that consist of concrete relations between people and groups.

There is validity to these possible methodological critiques, and future research utilizing alternative methodologies and data is needed to begin to address the important questions raised above. For example, rather than focusing on organizational claims-making during the course of political conflict, future research could utilize ethnographic and/or interview methods and data to begin to shine light on questions of what motivates activists to dedicate their time and resources to fighting the production of nuclear power. Ethnographic fieldwork within antinuclear organizations could begin to produce knowledge about the roles of race, class, and gender within antinuclear organizations that might impact our understandings of what constitutes energy justice. Similarly, conducting and transcribing in-depth interviews with antinuclear activists and organizational leaders could begin to answer questions about how experiences with broader systems of racism, sexism, and class domination compel activists into social movement activity. In-depth interviews would also reveal possible internal conflicts and/or divisions that shape and impact the claims the organization makes publicly, as well as any significant differences between members' beliefs and perceptions and those reflected in the organizations' political claims-making.

The present study critically analyzed the antinuclear power movement (or the movement *for* a carbon-free and nuclear-free energy future) in the United States using an environmental justice framework. It aimed to explore how different conceptualizations/discourses of social and environmental justice are constructed through the claims of social movement organizations on both levels of the movement. The goal was that through critical analyses of the claims made by the antinuclear power movement in the United States, as well as analyses of the historical/structural conditions these claims were made in response to, it would be possible to distill general principles of energy justice. The identification of general principles of energy justice, similar to the Principles of Environmental Justice, should guide future energy policy and the development of future energy systems to ensure social and environmental justice are maximized. This study thus represents a step in this process of theory and social justice building. As the preceding analysis has shown, more research utilizing multiple and complementary methodologies is needed in the future, not only on the antinuclear power movement but also on other movements for safe and clean energy systems. Humanity is entering the third decade of a century that so far has been characterized by rapidly increasing social inequalities (domestically and globally) and rapidly increasing climate change, both of which present serious and growing threats to the stability of all civilizations. Energy systems provide the connective tissue between these two existential threats. The connections between energy systems and social systems and between energy (in)justice and social (in)justice are plainly evident when we examine the claims made by groups fighting to highlight the social dimensions of energy systems, such as antinuclear groups and others. If humanity is to reverse the trends of widening social inequalities and increasing environmental and climatic disruptions, then we must not overlook these social dimensions of energy production, distribution, and consumption. In fact, advancing the cause of energy justice may prove to be an essential component of working toward broader goals of social and environmental justice. And advancing the cause of energy justice will require far more research from more disciplines utilizing more methodologies to advance as clear a picture as possible of what both just and unjust energy systems can look like in the future.

References

Arendt, Hannah. [1951] 1973. *The Origins of Totalitarianism*. New York: Harcourt.

Beyond Nuclear. February 3, 2010. "Press Release: Obama Administration Makes Major Moves to End Yucca Mountain Dumpsite Proposal". Takoma Park, MD: Beyond Nuclear. Retrieved February 4, 2010. (http://www.beyondnuclear.org/home/month/february-2010?currentPage=4).

———. 2012a. "Factsheet 'Loan Guarantees'". Takoma Park, MD: Beyond Nuclear. Retrieved February 15, 2012 (http://www.beyondnuclear.org/fact-sheets/).

———. January 26, 2012b. "Press Release: Beyond Nuclear Response to Publication of Report By DOE's Blue Ribbon Commission on America's Nuclear Future". Takoma Park, MD: Beyond Nuclear. Retrieved February 1, 2012 (http://www.beyondnuclear.org/radioactive-waste-whatsnew/2012/1/26/beyond-nuclear-response-to-publication-of-report-by-does-blu.html).

Blanding, Michael. 2012. "Vermont vs. Vermont Yankee: Should States Or The Rubber-Stamp NRC Decide Whether Nuclear Plants Continue To Operate?". *The Nation*. January 10, 2012.

Blau, Judith and Alberto Moncada. 2006. *Justice in the United States: Human Rights and the U.S. Constitution*. New York: Rowman and Littlefield Publishers.

Blue Ribbon Commission on America's Nuclear Future (BRC). 2011. *Report to the Secretary of Energy*. Washington, DC: BRC. Retrieved January 3, 2012 (http://cybercemetery.unt.edu/archive/brc/20120620220235/http://brc.gov/sites/default/files/documents/brc_finalreport_jan2012.pdf).

Blue Ridge Environmental Defense League (BREDL). 2011. "BREDEL Response to BRC's Draft Report". Retrieved Novemeber 2, 2011. (http://www.bredl.org/pdf3/111031_comments_on_draft_report_to_DOE_LZ.pdf).

Blum, Elizabeth. 2008. *Love Canal Revisited: Race, Class, and Gender in Environmental Activism*. Lawrence, KA: University Press of Kansas.

Bullard, Robert. [1990] 2000. *Dumping in Dixie: Race, Class, and Environmental Quality*. Boulder, CO: Westview Press.

———. 2005. "Environmental Justice in the Twenty-first Century". In Bullard, Robert (Ed.). *The Quest for Environmental Justice: Human Rights and the Politics of Pollution*. San Francisco, CA: Sierra Club Books, pp. 19–42.

Capek, Stella. 1993. "The 'Environmental Justice' Frame: A Conceptual Discussion and an Application". *Social Problems*. Vol. 40, No. 1, pp. 5–24.

Chen, Michelle. 2011. "The Radioactive Racism behind Nuclear Energy". *Color Lines*. March 23, 2011.

Cook, James. 1985. "Nuclear Follies". *Forbes*. February 11, 1985.

Donn, Jeff. 2011. "Safety Rules Loosened for Ageing Nuclear Reactors". *The Associated Press*. June 20, 2011.

Executive Order. January 29, 2010. "Memorandum For The Secretary Of Energy—Subject: Blue Ribbon Commission on America's Nuclear Future". Retrieved March 16, 2010 (http://cybercemetery.unt.edu/archive/brc/20120620220752/http://brc.gov/index.php?q=page/executive-order).

Foster, John Belamy, Brett Clark, and Richard York. 2011. *The Ecological Rift: Capitalism's War on the Earth*. Monthly Review Press.

Fraser, Nancy. 1997. *Justice Interruptus: Critical Reflections on the 'Postsocialist' Condition*. New York: Routledge.

———. 2000. "Rethinking Recognition". *New Left Review*. Vol. 3, (May–June), pp. 107–120.

———. 2001. "Recognition Without Ethics?" *Theory, Culture, and Society*. Vol. 18, pp. 21–42.

———. 2009. *Scales of Justice: Reimagining Political Space in a Globalizing World*. New York: Columbia University Press.

Fraser, Nancy and Axel Honneth. 2003. *Redistribution or Recognition? A Political-Philosophical Exchange*. New York: Verso.

Freudenburg, William and Robert Gramling. 1994. *Oil in Troubled Waters: Perception, Politics and the Battle over Offshore Drilling*. Albany, NY: SUNY Press.

———. 2012. *Blowout in the Gulf: The BP Oil Spill Disaster and the Future of Energy in America*. Cambridge, MA: MIT Press.

Friends of the Earth (FOE). July 15, 2010. "FOE Press Release: Did Democratic Leaders Try To Buy A House Seat with a $25 Billion Nuclear Bailout?". Washington, DC: FOE.

Giugni, Marco. 2004. *Social Protest and Policy Change: Ecology, Antinuclear, and Peace Movements in Comparative Perspective*. New York: Rowman and Littlefield.

Harney, Corbin. 2000. "Yucca Mountain: No Place for Nuclear Waste". Takoma Park, MD: NIRS. Retrieved February 15, 2011 (http://www.nirs.org/radwaste/yucca/yuccaltrbycorbin102400.htm).

Hawkesworth, Mary. 2006. *Feminist Inquiry: From Political Conviction to Methodological Innovation*. New Brunswick, NJ: Rutgers University Press.

Honneth, Axel. 1996. *The Struggle for Recognition: The Moral Grammar of Social Conflicts*. Cambridge, MA: MIT Press.

———. 2012. *The I in We: Studies in the Theory of Recognition*. New York: Polity.

International Business Times. 2010. "Budget Hawks Oppose Nuclear Loan Guarantees". Retrieved December 15, 2010 (http://www.ibtimes.com/budget-hawks-oppose-nuclear-loan-guarantees-250629).

Jacob, Gerald. 1990. *Site Unseen: The Politics of Siting a Nuclear Waste Repository*. Pittsburgh, PA: University of Pittsburgh Press.

Jasper, James. 1997. *The Art of Moral Protest: Culture, Biography, and Creativity in Social Movements*. Chicago: University of Chicago Press.

Johnston, Hank. 2002. "Verification and Proof in Frame and Discourse Analysis". In Klandermans, Bert and Suzanne Staggenborg (Eds.). *Methods of Social Movement Research*. Minneapolis, MN: University of Minnesota Press, pp. 62–91.

Joppke, Christian. 1993. *Mobilizing Against Nuclear Energy: A Comparison of Germany and the United States*. Berkeley: University of California Press.

Josephson Institute of Ethics. 2002. *Making Ethical Decisions*. Playa Del Rey, CA.

Kamps, Kevin. 2005. "Radioactive Racism: The History of Targeting Native American Communities with High-Level Atomic Waste Dumps". Takoma Park, MD: NIRS. Retrieved June 18, 2010 (http://www.nirs.org/radwaste/scullvalley/historynativecommunitiesnuclearwaste06142005.pdf).

Kuletz, Valerie. 1998. *The Tainted Desert: Environmental and Social Ruin in the American West*. London: Routledge.

Logan, John and Harvey Molotch. 1997. *Urban Fortunes: The Political Economy of Place*. Berkeley: University of California Press.

Macfarlane, Allison and Rodney Ewing, eds. 2006. *Uncertainty Underground: Yucca Mountain and the Nation's High-Level Nuclear Waste*. Boston: MIT Press.

Makhijani, Arjun. 2007. *Carbon Free and Nuclear Free: A Roadmap for U.S. Energy Policy*. Takoma Park, MD: Institute for Energy and Environmental Research (IEER).

Malm, Andreas. 2016. *Fossil Capital: The Rise of Steam Power and the Roots of Global Warming*. London: Verso.

McAdam, Doug. 1982. *Political Process and the Development of Black Insurgency, 1930–1970*. Chicago: University of Chicago Press.

McAdam, Doug, John McCarthy, and Mayer Zald (Eds.). 1996. *Comparative Perspectives on Social Movements: Political Opportunities, Mobilizing Structures, and Cultural Framing*. Cambridge, UK: Cambridge University Press.

McCarthy, John and Mayer Zald. 1977. "Resource Mobilization and Social Movements: A Partial Theory". *American Journal of Sociology*. Vol. 82, pp. 1212–1241.

Missourians for Safe Energy (MSE). January 31, 2011a. "MSE Letter to Missouri State Senators". Columbia, MO: MSE.

———. March 25, 2011b. "MSE Letter to State Legislatures". Columbia, MO: MSE.

Nuclear Information Resource Services (NIRS). 1998. "Press Release: 219 Groups Issue Petition Demanding Disqualification Of Yucca Mountain As Waste Site." Takoma Park, MD: NIRS. Retrieved March 10, 2011. (http://www.nirs.org/radwaste/yucca/yuccahome.htm).

———. December 2010a. "NIRS Newsletter". Takoma Park, MD: NIRS.

———. December 2010b. "NIRS List of Organization: Over 50 National & 700 State/Local Environmental & Public Interest Organizations Opposed To The

Yucca Mountain Nuclear Waste Dump". Takoma Park, MD: NIRS. Retrieved January 26, 2011. (http://www.nirs.org/radwaste/yucca/yuccaopponentslist.htm).

———. March, 24 2010c. "NIRS Principles for Safeguarding Nuclear Waste at Reactors". Takoma Park, MD: NIRS. Retrieved March 30, 2010 (https://www.nirs.org/radwaste/policy/hossprinciples3232010.pdf).

———. May 26, 2010d. "NIRS Action Alert: Your Actions Now Can Stop $9 Billion In New Nuclear Reactor Loans!" Takoma Park, MD: NIRS.

———. May 28, 2010e. "NIRS Action Alert: House Appropriations Committee Postpones Consideration Of Emergency Bill, $9 Billion In New Reactor Loans; But Don't Let Up Now! Keep Those Letters Pouring In!". Takoma Park, MD: NIRS.

———. June 16, 2010f. "NIRS Action Alert: Your Action Now Can Stop $9 Billion In New Nuclear Loans!". Takoma Park, MD: NIRS.

———. June 22, 2010g. "NIRS Action Alert: Correction!! House Appropriations Committee To Meet Thursday Afternoon—June 24—To ~~determine Fate Of $9 Billion~~ Consider $36 Billion In Taxpayer Loans For New Nuclear Reactors Last Chance To Act And Mobilize!". Takoma Park, MD: NIRS.

———. June 23, 2010h. "NIRS Action Alert: House Appropriations Subcommittee Abruptly Cancels Planned Thursday Afternoon Meeting; Still Time To Act To Stop $36 Billion In New Nuclear Loans!". Takoma Park, MD: NIRS.

———. July 1, 2010i. "NIRS Action Alert: Urgent Update House Appropriations Committee May Consider $9 Billion In Nuclear Loans Today Or Tomorrow". Takoma Park, MD: NIRS.

———. July 1, 2010j. "NIRS Press Release: Groups: Foreign Companies, Workers Are Big Early Winners Under Federal Loan Guarantees For Nuclear Reactors: *Taxpayer-Backed Nuclear Loan Guarantees Make July 4th 'Energy Non-Independence Day'; 2 of 3 Leading Loan Guarantees Candidates Have Major Foreign Ownership, Large Reactor Component Construction Jobs Are Overseas.*" Takoma Park, MD: NIRS.

———. July 2, 2010k. "NIRS Action Alert: House Approves Emergency Funding Bill W/$9 Billion In Nuclear Loans Bill Will Now Go Back To Senate". Takoma Park, MD: NIRS.

———. July 23, 2010l. "NIRS Action Alert: Good News From Washington (For A Change!). Senate Rejects Emergency Funding Bill With $9 Billion In Nuke Loan Guarantees. Senate Drops Nuke-Larded Climate Bill And Senate Committee Cuts House Committee's Nuke Loan Program". Takoma Park, MD: NIRS.

———. July 30, 2010m. "NIRS Action Alert: Sen Hutchison Threatens New Attempt To Give $9 Billion In New Nuke Loans As Top Projects Teeter Toward Collapse. Your Actions Now Can Stop Them Again!". Takoma Park, MD: NIRS.

———. December 7, 2010n. "NIRS Action Alert: Urgent! DOE Yet Again Seeking $9 Billion For New Reactor Loans For The Last Time This Year—Tell Congress: No Taxpayer Bailouts For Giant Nuclear Corporations!". Takoma Park, MD: NIRS.

———. December 23, 2010o. "NIRS Action Alert: Happy Holidays—Another Huge Victory! Government Funding Bill Enacted Without Nuclear Loans!". Takoma Park, MD: NIRS.

———. February 15, 2011a. "NIRS Action Alert: Obama Fy 2012 Budget: Cut Heating Assistance For Poor, Cut Clean-Up Of Great Lakes... But Shower Billions Of Dollars In Loans To Wealthy Nuclear Utilities And Foreign Reactor Manufacturers! Let's Get To Work....Tell Congress: No Way!". Takoma Park, MD: NIRS.

———. February, 16 2011b. "NIRS Action Alert: Tell Missouri Lawmakers: Oppose Sb 50 And Hb 124—Ratepayers Must Not Pay For New Nuke Costs In Advance". Takoma Park, MD: NIRS.

———. March 3, 2011c. "NIRS Action Alert: Tell Congress Now The American People Say: Don't Just Prevent New Nuclear Subsidies, Roll Back The Existing Ones!". Takoma Park, MD: NIRS.

———. March 7, 2011d. "NIRS Action Alert: Make Sure Congress Hears Your Voice! End The Nuclear Reactor Loan Program!". Takoma Park, MD: NIRS.

———. April 13, 2011e. "NIRS Action Alert: Organizational Sign-On Letter To End Nuclear Loan Program". Takoma Park, MD: NIRS.

———. May 25, 2011f. "NIRS Organizational Sign-On Letter". Retrieved May 25, 2011 (http://www.nirs.org/neconomics/lgsignon52511.pdf).

———. January 25, 2012a. "NIRS Press Release: Environmental Groups, 5,000+ People Tell Secretary Chu: Reject The BRC Report—Do Not Use Closed Reactors As Excuse To Create Temporary Storage Site And Initiate Mass Transportation Of Radioactive Waste Across The U.S.". Takoma Park, MD: NIRS.

———. January 25, 2012b. "NIRS Talking Points for BRC Public Meetings on Radioactive Waste Policy". Takoma Park, MD: NIRS.

———. January 25, 2012c. "NIRS Letter to Secretary of Energy". Takoma Park, MD: NIRS. Retrieved February 1, 2012 (http://www.nirs.org/radwaste/policy/chuwasteletter01242012.pdf).

Nye, D. 1998. *Consuming Power: A Social History of American Energies*. Cambridge, MA: MIT Press.

Pellow, David. 2007. *Resisting Global Toxics: Transnational Movements for Environmental Justice*. Cambridge, MA: MIT Press.

Pellow, David and Robert Brulle. 2005. "Power, Justice, and the Environment: Toward Critical Environmental Justice Studies". In Pellow, David and Robert Brulle (Eds.). *Power, Justice, and the Environment: A Critical Appraisal of the Environmental Justice Movement*. Cambridge, MA: MIT Press, pp. 1–19.

People's Campaign to Close Vermont Yankee. August 15, 2011. "Open letter: A Call To Join The Organizing Of A Regional Campaign Of Nonviolent Direct Action To Shut Down Vermont Yankee". Brattleboro, VT: People's Campaign to Close Vermont Yankee.

Perkins, John. 2017. *Changing Energy: The Transition to a Sustainable Future*. Berkeley: University of California Press.

Physicians for Social Responsibility (PSR). 2011. "Physicians for Social Responsibility's Comments on the BRC's Draft Report". Washington, DC: PSR.

Piven, Frances Fox and Richard Cloward. 1979. *Poor People's Movements: Why They Succeed and How they Fail.* New York: Vintage Books.

Schell, Jonathan. 2007. "The Spirit of June 12". *The Nation*. July 2, 2007.

Schlosberg, David. 2007. *Defining Environmental Justice: Theories, Movements, and Nature*. New York: Oxford University Press.

Scott, Rebecca. 2010. *Removing Mountains: Extracting Nature and Identity in the Appalachian Coalfields*. Minneapolis: University of Minnesota Press.

Sewell, William. 1980. *Work and Revolution in France*. Cambridge, UK: Cambridge University Press.

Shut It Down Affinity Group. January 20, 2012. "Shut It Down Affinity Group Press Release: Women Arrested at Vermont Yankee While Mourning Cancer Deaths". Brattleboro, VT: Shut It Down Affinity Group.

Snow, David and Robert Benford. 1986. "Frame Alignment Processes, Micro-Mobilization and Movement Participation". *American Sociological Review*. Vol. 51, pp. 464–481.

Snow, David and Robert Benford. 1992. "Master Frames and Cycles of Protest" In Aldon Morris and Carol Mueller (Eds.). *Frontiers in Social Movement Theory*. New Haven, CT: Yale University Press.

Somers, Margaret. 2008. *Genealogies of Citizenship: Markets, Statelessness, and the Right to Have Rights*. Cambridge, UK: Cambridge University Press.

Speth, James. 2009. *The Bridge at the Edge of the World: Capitalism, the Environment, and Crossing from Crisis to Sustainability*. New Haven, CT: Yale University Press.

Szasz, Andrew. 1994. *Ecopopulism: Toxic Waste and the Movement for Environmental Justice*. Minneapolis: University of Minnesota Press.

Sze, Julie. 2005. "Race and Power: An Introduction to Environmental Justice Energy Activism". In Pellow, David and Robert Brulle (Eds.). *Power, Justice, and the Environment: A Critical Appraisal of the Environmental Justice Movement*. Cambridge, MA: MIT Press. pp. 101–115.

———. 2007. *Noxious New York: The Racial Politics of Urban Health and Environmental Justice*. Cambridge, MA: MIT Press.

Tarrow, Sidney. 1994. *Power in Movement*. Cambridge, UK: Cambridge University Press.

Tatum, J. 2000. *Muted Voices: The Rediscovery of Democracy in the Shaping of Technology*. Bethlehem, PA: Lehigh University Press.

Taylor, Dorceta. 2000. "The Rise of the Environmental Justice Paradigm: Injustice Framing and the Social Construction of Environmental Discourses". *American Behavioral Scientist*. Vol. 43, No. 4, pp. 508–80.

U.S. Energy Information Administration. 2021. *Levelized Costs of New Generation Resources in the Annual Energy Outlook 2021*. Retrieved July 15, 2021 (https://www.eia.gov/outlooks/aeo/pdf/electricity_generation.pdf).

Vandenbosch, Robert and Susanne Vandenbosch. 2007. *Nuclear Waste Stalemate: Political and Scientific Controversies*. Salt Lake City: University of Utah Press.

Van Doren, Peter. 2010. "Stop Nuclear Welfare". *New York Times*. Retrieved February 16, 2010 (http://roomfordebate.blogs.nytimes.com/2010/02/16/a-come-back-for-nuclear-power/?_php=true&_type=blogs&_php=true&_type=blogs&_r=1&).

Van Gerven, Jesse. 2014. "It Is Laced With Faults: American Indians, Public Participation, and the Politics of Siting a High-Level Radioactive Waste Repository". *Societies Without Borders*. Vol. 9, No. 2. pp. 161–187.

Vermont Citizens Action Network (VTCAN). 2010. "VTCAN Fact Sheet: Where Does Vermont Yankee's Waste Go?". Brattleboro, VT: VTCAN. Retrieved March 30, 2011 (http://www.vtcitizen.org/waste.shtml).

Walker, Samuel. 2009. *The Road to Yucca Mountain: The Development of Radioactive Waste Policy in the United States*. Berkeley: University of California Press.

Walsh, Edward. 1981. "Resource Mobilization and Citizen Protest in Communities around Three Mile Island". *Social Problems*. Vol. 29, No. 1, pp. 1–21.

Wright, Erik Olin. 2010. *Envisioning Real Utopias*. New York: Verso.

Zald, Mayer and John McCarthy (Eds.). 1987. *Social Movements in an Organizational Society: Collected Essays*. New Brunswick, NJ: Transaction Press.

Index

Abalone Alliance's protests, 104–5
ACT. *See* Association of Concerned Taxpayers (ACT)
Action Alert, NIRS, 37–39, 41–42, 46–50, 52, 100–101
AEC. *See* Atomic Energy Commission (AEC)
African American communities, 94, 107
AGE. *See* Alliance for Generational Equity (AGE)
Alexander, Lamar, 47
Alliance for Generational Equity (AGE), 44–47
AmerenUE, 98–99
antinuclear power movement, 3, 7–9, 21, 27, 93; basic concerns, 84–85; BRC and, 83–85; coalition for, 101; people power and, 100–101; for radioactive waste management, 57–92; VTCAN links, 105–8
antinuclear SMOs, 4–5, 9, 20, 29–31, 75, 104–15
Arendt, Hannah, 25
Association of Concerned Taxpayers (ACT), 44–47
atom, 9–10
Atomic Energy Act, 4, 57, 64–65
Atomic Energy Commission (AEC), 4–5, 57–58, 89
"Atoms for Peace" program, 4, 58
Ayers, Mark, 66

Bailey, Vicky A., 66
Bear, Lawrence, 72–73
Bear, Richard, 72–73
Berry, Marion, 42
Beyond Nuclear, 2, 5, 35
BIA. *See* Bureau of Indian Affairs (BIA)
Blue Ribbon Commission (BRC), 2, 7, 9, 31, 63, 66, 74–75, 82, 87, 90
Blue Ridge Environmental Defense League (BREDL), 74
boiling water reactors (BWR), 12–13
Bradford, Peter, 102
BRC. *See* Blue Ribbon Commission (BRC)
BREDL. *See* Blue Ridge Environmental Defense League (BREDL)
A Brighter Tomorrow: Fulfilling the Promise of Nuclear Energy (Domenici), 66
Bullard, Robert, 19, 123
Bureau of Indian Affairs (BIA), 70, 73, 90
BWR. *See* boiling water reactors (BWR)

Calvert Cliffs project, Maryland, 43
CAN. *See* Citizens Awareness Network (CAN)
CANT. *See* Citizens Against Nuclear Trash (CANT)
Capek, Stella, 17–19
Carnesale, Albert, 66
Chernobyl disaster (1986), 6
Chu, Steven, 2–3, 60, 63, 85–86
Citizens Against Nuclear Trash (CANT), 94–95
Citizens Awareness Network (CAN), 1–2
Citizens Environmental Coalition of Albany, 2
Clamshell Alliance's protests, 104–5
Clean Water Action, 52
Clyburn, Whip James, 42
cognitive liberation, 16
collective action frames, 17
Competitive Enterprise Institute, 48
construction work in progress charges (CWIP), 21, 28, 95–104, 120, 125, 131
Continuing Resolution, 48–50
Crowe, Frances, 1
CWIP. *See* construction work in progress charges (CWIP)

data sources, 29–33
Defining Environmental Justice: Theories, Movements, and Nature (Schlosberg), 25–26
Department of Energy (DOE), 6–7
Department of Energy Organization Act (1977), 6
discourse, 29–30
distributive (in)justice: of federal "loan guarantee" program, 21, 35–56; local financing and, 103–13; with nuclear waste policy, 80, 89–91, 95, 100; of public financing, 18, 120–24; theories, 21–26, 129–35
DOE. *See* Department of Energy (DOE)
Domenici, Pete V., 66

economic resources, 16
Ecopopulism: Toxic waste and the Movement for Environmental Justice (Szasz), 64
EDF. *See* Electricite de France (EDF)
Edwards, Chet, 42
Edwards, Donna, 42
Eisenhower, Susan, 66–67
EJM. *See* environmental justice movement (EJM)
Electricite de France (EDF), 48
emergency supplemental funding bill, 36–41
energy justice, potential principles of, 119–29; energy systems, development and implementation of, 125–27; federal "loan guarantee" program, multidimensional injustices with, 35–55, 119–20; future environmental injustices, 124; independent regulatory, 127; public financing of energy systems, 120–24; radioactive waste issue, analyses of, 124–25
Energy Policy Act (2005), 6, 47–48
Energy Reorganization Act (1974), 5, 58
Entergy Corporation, 109–10
environmental justice, 17–29, 129–35; conceptualization of, 83; definition, 82–83; energy activism, 26, 48–49, 57; framework of, 108–9
environmental justice movement (EJM), 18–20, 26
environmental racism, 18–20, 94, 107, 123
ethical decision-making, 136
ethics, 135–37
Executive Order 12898, 19

Fattah, Chaka, 42
federal financing of new nuclear reactor construction, 35–55; round one (April and May 2010), 36–41; round two (June and July 2010), 41–47;

round three (June and July 2010), 47–48; round four (December 2010), 48–50; round five (February–June 2011), 50–55
federal "loan guarantee" program, 6, 8, 18, 21, 35–55, 119–20; maldistributive aspects of, 36–38, 42–50, 55; multidimensional injustices associated with, 21, 27–29, 36, 41–47, 54, 57, 99–102, 113; recognitional justice, 50–55; representative justice, 50–55
Fermi, Enrico, 4
fission, 10–11
folk paradigms, 22
frame alignment, 17
framing, 16–17
Fraser, Nancy, 22–24, 27, 57, 66, 68, 90
Friends of the Earth, 52
Fukushima disaster, 6, 17, 50–53, 79, 102, 105

Genealogies of Citizenship: Markets, Statelessness, and the Right to Have Rights (Somers), 21
Graham, Lindsey, 6
Graves, Ellen, 1
Great East Japan Earthquake (2011), 6
Greenpeace, 52

Hagel, Chuck, 67
Hamilton, Lee, 66
HANDS. *See* Humans Against Nuclear-Waste Dumping (HANDS)
Hatch, Orin, 47
hazardous materials, 64–65
Hegelian identity politics, 23
Hegelian recognition schema, 23
Heritage Foundation, 48
Herman, David, 44
high-level radioactive waste (HLRW), 13–15, 28, 57–92; BRC recommendations for management, 76–80 (consent-based approach, 76–77; consolidated interim storage facilities, development of, 78–79; new organization establishment, 77; new permanent geologic disposal facility, development of, 78; utility waste disposal fees, 77–78); deep geologic disposal sites for, 91; hazardous material production and disproportionate burdens link, 82–83; management, 57–92; on Native Americans land, 69–72; NIRS arguments on, 84–85; Principles of HOSS and, 86–88; report, 75–76; storage of, 69–72
high-level radioactive waste repository (HLRWR), 14–15
HLRW. *See* high-level radioactive waste (HLRW)
HLRWR. *See* high-level radioactive waste repository (HLRWR)
Honneth, Axel, 22
House Appropriations Committee, 37
Hoyer, Steny, 42
Humans Against Nuclear-Waste Dumping (HANDS), 71
Hydrology Guideline, 61

identity politics, 23
innovation, 135–37
intergenerational equity, 45

Jones, Gordon, 44
justice, 26. *See also* environmental justice; social justice

Kamps, Kevin, 69–70, 74, 132–33
Kuletz, Valerie, 68

Lash, Jonathan, 67
Laws, Rufina Marie, 71
LCA. *See* life-cycle assessment (LCA)
LCOE. *See* levelized costs of electricity (LCOE)
Leroy, David, 70–71
LES. *See* Louisiana Energy Services (LES)

levelized costs of electricity (LCOE), 40
life-cycle assessment (LCA), 39–40
local financing of new nuclear reactor construction, 93–115; CWIP financing for, 96–104; and safety of nuclear facilities, 104–15
local opposition, 3–4, 70, 93–115, 137
Louisiana Energy Services (LES), 94–95

Macfarlane, Allison, 67
maldistribution, 24–25, 28, 55, 84, 89, 92, 106, 112; of benefits and burdens, 112, 131–34; claims regarding, 50; of financial risks, 38; injustices of, 70, 129–35; problems of, 45–46
Manhattan Project, 31, 57
Mariotte, Michael, 43
McAdam, Doug, 16–17
McCarthy, John, 15
McConnell, Mitch, 47
Mescalero Apache Tribe, 71–72
Mescalero Tribal Council, 71–72
Meserve, Richard A., 67
meta-injustice, 25
methodology, 29–33
misrecognition, 23–25, 44–45, 55, 89, 91–92, 95; injustice of, 70, 129–35; problems of, 55; of public concerns and interests, 40, 84, 86, 91–92, 98, 103, 106, 112, 130–34; public financing and, 28, 120–24; status model of, 23
misrepresentation/misframing, 24–25, 28, 40, 42–45, 55, 63, 66, 68–70, 86, 89, 90, 92, 95, 106, 130–34; ordinary, 25, 66, 68–69, 90, 133; political injustices of, 68–69, 129–35
Missourians for Safe Energy (MSE), 30, 97–102
Moniz, Ernie, 67–68
moral philosophy, 15–29
MSE. *See* Missourians for Safe Energy (MSE)

Nader, Ralph, 58
National Environmental Justice Advisory Council (NEJAC), 19
National Taxpayers Union, 48
Native American communities, 70
NEI. *See* Nuclear Energy Institute (NEI)
NEIS. *See* Nuclear Energy Information Services (NEIS)
NEJAC. *See* National Environmental Justice Advisory Council (NEJAC)
Nestel, Hattie, 1
neutrons, 9–10
new nuclear reactor construction: CWIP financing for, 96–104; federal financing of, 35–55; local financing of, 93–115
NIRS. *See* Nuclear Information and Resource Services (NIRS)
Non-proliferation Policy Education Center, 48
Noranda Aluminum Holding Corporation, 101
NRC. *See* Nuclear Regulatory Commission (NRC)
Nuclear Energy Information Services (NEIS), 2
Nuclear Energy Institute (NEI), 6
nuclear facilities safety, 104–15
nuclear-industrial-academic-complex, 68
Nuclear Information and Resource Services (NIRS), 2, 5, 8, 35, 37–59, 61, 66, 81–87, 100–101
nuclear power, 1–33; consumption of, 57, 95, 117–18, 138, 142; distribution, 18–30, 36, 39, 43, 46, 52–57, 76, 80–84, 94–114, 117–33; energy justice, potential principles of, 119–29; environmental justice and, 17–29; ethics, 135–37; innovation, 135–37; local opposition to, 3–4, 57–92; new nuclear reactor construction(federal financing of, 35–55; local financing of, 93–115); Obama administration

on, 7–8; overview of, 9–15; plants, 12; politics of, 4; production, 1–21, 27–28, 31, 36, 40–45, 54, 57, 64–69, 75–76, 81–92, 95, 103–5, 107, 110, 114–15, 117–33, 138, 141; social movement studies and, 15–29; in the United States, 3–33
Nuclear Regulatory Commission (NRC), 5–7, 30, 37, 58, 60, 63, 74, 79, 89, 94–95, 102, 105–6, 109–10, 113–14, 125, 131–32
nuclear renaissance, 6–8
Nuclear Resource Group's project in Southwest, TX, 37–38
Nuclear Waste Policy Act Amendments (NWPAA, 1987), 7, 59
Nuclear Waste Policy Act (NWPA, 1982), 58–59, 61
nucleus, 9

Obey, David, 42
Olson, Mary, 66
ordinary misrepresentation, 25, 66, 68–69, 90, 133

Pellow, David, 17
Peterson, Per, 68
Physicians for Social Responsibility (PSR), 84–85
Plutonium-239, 13
political process models, 16
pressurized water reactors (PWR), 12–13
principle of energy justice, 91, 119–29
Principles for Safeguarding Nuclear Waste at Reactor Sites, 3
Principles of Hardened On-Site Storage (HOSS), 86–88
Private Fuel Storage LLC (PFS), 72
problem of displacement, 139
protons, 9–10
PSC. *See* Public Service Commission (PSC)
PSR. *See* Physicians for Social Responsibility (PSR)
Public Service Commission (PSC), 97–98
PWR. *See* pressurized water reactors (PWR)

Quintana, Danny, 72–73

radiation, 14–15
radioactive racism, 69, 74, 76–77
radioactive waste, 2–9, 13–15, 93; handling of, 4–5; local opposition to, 3–4; management, 57–92; Obama administration on, 7–8; politics of, 4; production of, 3, 13–15; public protest over, 3–33, 58–92; storage of, 2–3
Radioactive Waste Policy, 30
recognition (in)justice, 22–25, 41, 50–55, 76, 84, 89, 95, 101–4, 129–35; antinuclear power movement, 46, 50–55; in democratic practices, 23–24; issues of, 22–27; local level antinuclear organizations and, 28, 95–115; in political process, 19; radioactive waste management and, 57–92; redistribution to, 22–23; social justice paradigms of, 36
redistribution, 22–25
Redistribution or Recognition? A Political-Philosophical Exchange (Fraser and Axel), 22
regional equity, 59
Reid, Harry, 62
representational (in)justice, 68–69, 81–82, 84, 90–92, 95, 98, 105–6, 119, 129–35; antinuclear power movement and, 46, 50–55; CANT and, 94–95; definition, 24–25; local antinuclear organization and, 95–115; local level antinuclear organizations and, 28; of misframing, 25, 63, 68–69, 81, 90, 133; political categories of, 118; in public deliberations, 24; radioactive waste management and, 57–92; social justice paradigms of, 24–27, 36; theoretical interests, 30, 129–35

Resource Conservation and Recovery Act (RCRA, 1976), 64
"Resource Mobilization and Social Movements: A Partial Theory" (McCarthy and Zald), 15
resource mobilization theory, 15–17
Rowe, John, 68

Scales of Justice: Reimagining Political Space in a Globalizing World (Fraser), 23–24
Schlosberg, David, 25–27
Scowcroft, Brent, 66
Screw Nevada Bill (1987), 59
Seabrook Nuclear Power Plant, 104
Sharp, Phil, 68
Shelburne Falls, 1
Shelby, Richard, 47
Sierra Club Legal Defense, 94
Site Suitability Guidelines, 61
Skull Valley Band of Goshutes, 71–73
SMOs. *See* social movement organizations (SMOs)
social justice, 25–29
social movement organizations (SMOs), 4, 15–18, 30–33
social movement studies, 15–29; cognitive liberation, 16–17; collective action frames, 17; frame analysis in, 17; organization in community, level of, 16–17; political environment, changing opportunities in, 16–17; resource mobilization theory, 15–17
social movement theories, 15–29
Somers, Margaret, 21, 26, 141
sound radioactive waste disposal policy, 63
status model of misrecognition, 23
Strontium-90, 13
Sze, Julie, 26, 123

Tarrow, Sidney, 16
taxpayers, 49
Taxpayers for Common Sense, 48
Taylor, Dorceta, 17
Tea Party movement, 45, 47–48
Tennessee Valley Authority (TVA), 84
Thorpe, Grace, 71, 81
Three Mile Island accident (1979), 5–6, 102, 104–5
Toshiba Corporation, 43, 48
Treaty of Ruby Valley (1863), 59
TVA. *See* Tennessee Valley Authority (TVA)

UCS. *See* Union of Concerned Scientists (UCS)
Uncertainty Underground: Yucca Mountain and the Nation's High-Level Nuclear Waste (Macfarlane), 67
Union Electric, 97, 99
Union of Concerned Scientists (UCS), 110
Unistar's project at Calvert Cliffs, MD, 37–38
uranium, 9–15
uranium mining communities, 3

Vermont Citizens Action Network (VTCAN), 106–8
Vermont Yankee Nuclear Reactor, 1, 105–12
Vernon, Vermont, 1
voluntary siting, 69–70
VTCAN. *See* Vermont Citizens Action Network (VTCAN)

Walsh, Edward, 15–16
Wieland, Paki, 1

Yankee Rowe Nuclear Power Plant, Vermont, 105–12
YMP. *See* Yucca Mountain Project (YMP)
Yucca Mountain Project (YMP), 7, 14, 29, 57–92
Yucca Mountain Repository, 7, 14

Zald, Mayer, 15

About the Author

Jesse P. Van Gerven has been a lecturer in the Science, Technology, and Environmental Studies program at Butler University since 2014. He regularly teaches courses in the areas of the social studies of science and technology and environmental studies. His research focuses on energy policy, food systems, local-sustainable agriculture, and social movements for energy and food justice. He has published in the areas of race and energy policy and critical methodologies. He lives, works, and plays in the Indianapolis area with his wife, two sons, and cat and dog.

www.ingramcontent.com/pod-product-compliance
Ingram Content Group UK Ltd.
Pitfield, Milton Keynes, MK11 3LW, UK
UKHW020842290926
15470UKWH00004B/1493

9 781793 620453